Productivity Improvements Through TPM

The Philosophy and Application of Total Productive Maintenance

R. K. Davis

Prentice Hall
New York London Toronto Sydney Tokyo Singapore

First published 1995 by
Prentice Hall International (UK) Limited
Campus 400, Maylands Avenue
Hemel Hempstead
Hertfordshire, HP2 7EZ
A division of
Simon & Schuster International Group

Typeset in 10/12pt Times
by Hands Fotoset, Leicester

Printed and bound in Great Britain by
T.J. Press (Padstow) Ltd

Library of Congress Cataloging-in-Publication Data

Davis, R. K. (Roy K.)
Productivity improvements through TPM : the philosophy and application of total productive maintenance / R.K. Davis.
p. cm. — (The Manufacturing practitioner series)
Includes index.
ISBN 0-13-133034-9
1. Total productive maintenance. 2. Plant maintenance.
I. Title. II. Series.
TS192.D38 1994
658.2′02—dc20 94–31925
CIP

British Library Cataloguing in Publication Data

A catalogue record for this book is available from the British Library

ISBN 0-13-133034-9

1 2 3 4 5 99 98 97 96 95

This book is dedicated to all of the TPM Champions out there in industry who, in spite of the difficulties involved, are determined to keep pushing onwards towards the better tomorrow that they know TPM will bring for their businesses and everyone that works within them. I salute you and hope that this book in some way will help you to further the cause.

Contents

Preface

I first encountered TPM in 1989 when I was carrying out research for a large manufacturing organisation into what was 'best practice' in the maintenance of production machinery. I had been given the task of 'getting their act together' with regard to the maintenance and the general state of the manufacturing facilities within the group. As I travelled around the various operating divisions, made contacts outside of the group, attended training courses and seminars, listened to operations and maintenance people, I encountered two main recurring themes to their problems, namely:

1. The status of the maintenance function within most operating companies was not very high. Maintenance was regarded as a necessary evil, an overhead, an area where considerable amounts of money were spent but little return could be seen.
2. Maintenance personnel would like to carry out more planned preventive or predictive maintenance, but they did not have the time or resources to do so due to the heavy workload which was mainly composed of fixing machinery that had broken down.

It became clear to me that whatever sophisticated tools and techniques were introduced into the maintenance function, they would not help unless these two fundamental problems could be resolved.

When I was first introduced to TPM in its purest Japanese form, I found it a little confusing and wondered if it could work in countries other than Japan, with its unique traditions and culture. I found that the texts available explained the principles of TPM and how it is applied in Japan, but did not answer many of my questions and, in particular, did not describe how to introduce and implement TPM in a typical Western business. Further investigation, including a trip to Japan to see TPM in action, helped to clarify my thoughts and prove to me that TPM could work elsewhere, and that it provided the answer to the two fundamental problems that my research had identified. I then set about developing the TPM approach to suit the operating companies that I was at that time assisting, by tailoring training packages, implementing pilot areas and learning a great deal about how you actually apply TPM on different sites.

In the early days I made a lot of mistakes and had to redesign my implementation approach quite considerably until it developed into a robust, practical and structured way of implementing TPM, which worked in almost any industry sector and also in various areas of an operating business. In the light of my experiences and also because of my belief that there is a need for a text of this kind, I felt it necessary to write this book which is both an introduction to the fundamentals of TPM and also a guide to the successful implementation of TPM.

Chapter 1 provides an introduction to the subject and its historical background, whereas Chapter 2 explains the role of TPM and where it fits in with other 'best practice' and total quality initiatives which the reader may have encountered. The general principles of TPM, its main elements and scope of application are covered in Chapter 3, along with a detailed explanation of the six big losses and overall effectiveness. The tangible and intangible benefits of TPM are explained in Chapter 4, including the use of overall effectiveness to justify expenditure. Chapter 5 deals with the practicalities of introducing TPM to a business, getting started and dealing with people at all levels before discussing the overall business implications of TPM.

Throughout the book I have tried to show how TPM can be applied to almost any operating company in any industry sector and also to most areas of that company. It has been necessary, however, to emphasise the application of TPM in production areas, as this is usually of most importance to businesses initially and, therefore, the book mostly refers to machinery. The reader should bear in mind that the same principles and practices do apply to most other facilities and are not exclusive to production machinery.

The contents of the book are based upon my practical experience of implementing TPM over the last four years, and explain the TPM approach and understanding, which developed through working with many teams in different industry sectors and business areas during that time.

If the reader is not aware of the benefits that TPM can bring, then I hope that this book will provide inspiration. If the reader is already a 'disciple' of the TPM philosophy, then I hope that the book will provide help and encouragement.

Acknowledgements

I would like to acknowledge the support and encouragement provided by my wife, the help and assistance provided by colleagues at the TPM Centre of the Productivity Improvements Division of Monition Limited and also, the UK Department of Industry who provided valuable support for my early research into the application of TPM and traditional maintenance techniques.

1 Introduction to Total Productive Maintenance

Total Productive Maintenance (TPM) is the Japanese approach to maximising the effectiveness of the facilities that we use within our businesses. It not only addresses maintenance but all aspects of the operation and installation of those facilities, and at its very heart lies the motivation and enhancement of the people who work within the company.

The three components of TPM are as follows:

1. Total approach. An all-embracing philosophy which deals with all aspects of the facilities employed within all areas of an operating company and the people who operate, set up and maintain them.
2. Productive action. A very pro-active approach to the condition and operation of facilities, aimed at constantly improving productivity and overall business performance.
3. Maintenance. A very practical methodology for maintaining and improving the effectiveness of facilities and the overall integrity of production operations.

The maintenance component of TPM is not that which is traditionally recognised by operating companies. This conjures up the vision of a maintenance fitter, dripping in oil and grease, grasping a spanner and delving into the entrails of a broken machine. It is much more related to the maintenance of the integrity of manufacturing operations and pro-active (rather than reactive) maintenance which concentrates on all aspects of the condition and operation of operating facilities.

The essence of TPM is teamwork, focused on the condition and performance of particular facilities. The team is composed of people who operate, set up and maintain the facilities with, in some instances, the addition of people who are involved in the provision of planning or engineering support to the facilities.

This team may be different from the teams previously seen in organisations. These may have been set up to solve specific problems or to carry out a

The TPM team

special project and will often have included people from a variety of functions and disciplines. What makes TPM work is the team, and what makes the team work is the fact that they are focused on *their* facilities, *their* everyday problems and *their* environment.

TPM was first introduced in Japan about 20 years ago and has been applied rigorously in many Japanese factories, particularly in the past 10 years. The planning and implementation of TPM in Japanese factories have been supported by an influential maintenance institute which each year awards a 'PM' prize to the best Japanese (and more recently, overseas) companies in recognition of their achievements in the application of TPM. The prize has been awarded to companies in various industry sectors including the following:

- Chemicals
- Foods
- Rubber
- Metals
- Automotive
- Glass

This seems to confirm that TPM as an approach is applicable across most manufacturing industry sectors. The first English texts on the subject were published around 1987 and since then, the business interest in TPM has gradually increased.

There is a tendency to regard TPM as an approach that is only applicable to the high- to medium-volume production areas of operating businesses, and a great deal of publicity is given to the application of TPM in these areas. This is probably because the results of TPM implementation in volume production areas can be quite spectacular and provide very substantial benefits to a business. On the contrary, TPM should not be confined to high- to medium-volume production areas and also should not be regarded as only applicable to production machinery on the factory floor. It is a philosophy which should permeate throughout an operating company and touch people at all levels in the organisation.

TPM has been successfully applied in low-volume production, high- to low-volume assembly (automated and hand assembly), test and development areas, offices, warehouses, etc. and across the whole range of industry sectors, with different company sizes and types. Within a business which is implementing TPM, no employee should remain untouched by some aspect of TPM, although it will normally be those working in the most significant production areas who are initially involved.

TPM achieves substantial improvements in the effectiveness of facilities, the working environment, people's morale and overall business performance through the gradual implementation of many small improvements and brings about a process of 'creeping change' which is sometimes not easily perceived by those working within the organisation. The pace of that change and the corresponding level of benefits achieved will vary from business to business and from site to site. This is due to the fact that the operating conditions, traditional values and culture, people skills and experience, facilities, technology and organisational structure will usually differ to some extent. The philosophy and principles of TPM will not vary, but the way in which it is introduced and implemented will need to be tailored to suit every application.

TPM will be introduced by the senior managers of a business, who may decide to bring in some expert assistance to provide advice, training and support, but it is the team members who implement TPM through the many and detailed activities that they carry out.

2 The Role of TPM

When first introduced to TPM, its principles and approach it can be difficult to understand what role TPM can play within a company and how it fits in with other concepts and techniques. Questions such as the following will probably arise:

- How can it work in my business?
- How does TPM fit in with total quality?
- Is it part of just-in-time philosophy?
- Does it replace other maintenance techniques?
- Is it just another Japanese 'buzz word'?
- When is the right time to start TPM?

It is important that we clearly understand the role of TPM and are able to provide answers. This chapter will explain the role of TPM and its relationship to other business approaches and maintenance practices, but first let us explore the situation and attitudes which prevail in a typical operating business.

2.1 The typical scenario

A look at an operating company that is not employing TPM or TPM-type activities will reveal a very typical scenario, namely:

- The incidence of machinery and equipment breakdowns is high, causing frequent disruption of production schedules.
- The time taken to change from one product type to another product type is excessive and causes significant lack of availability.
- The quality of products is inconsistent, often requiring high levels of inspection.
- Employee morale is low, not just at factory floor level but throughout the organisation.

Demotivated workforce

- There is a blame culture where departments blame each other for problems, and relationships are often strained.
- Managers are constantly reacting to crises of one sort or another and do not feel that they are in control of the situation.
- Individuals have become quite adept at reacting to panics and often operate under high levels of stress.
- The working environment is dirty, untidy and generally not very good.

Many of the Production Managers who I know are constantly under stress, work long hours, often receive panic telephone calls at home and have very little time to plan for the long or even medium term. Their working day consists of a series of problems (or challenges, as the theorists urge us to regard them), which have to be resolved quickly, punctuated by a badly run meeting or two and some paperwork. They are operating in reaction mode and end the day exhausted, with seemingly very little to show for their efforts.

I am sure that they would really like to 'walk the floor', as all good managers should, spending time with factory floor personnel. They would be delighted to be able to plan and support improvement initiatives and to learn about new approaches and techniques, but the reactive pressures are so great that they do not feel that they can do what they know they should be doing. The sad fact is that this situation will not alter by itself and panic mode will continue to prevail.

What might also surprise and sadden many managers is the way in which the factory floor level personnel perceive them. To many, their supervisors, foremen and managers are incompetent, badly informed and constantly making decisions which they cannot understand. It is precisely because of the regime of crisis management, lack of management visibility, poor communications and lack of planning that this perception exists and management credibility is all too often very low indeed. From the factory floor point of view, many of the decisions that are taken support the perception of an incompetent management, as the reasons for the decisions are not communicated or understood. Factory personnel feel that they are never consulted or listened to, even on matters upon which they are eminently qualified to advise. Thus they tend to let 'management' get on with making ill-informed decisions and mistakes and become more and more frustrated.

It is only by introducing and supporting TPM that a 'better tomorrow' can be achieved. It takes hard work and requires persistent effort, but because it is based upon common sense and a sound philosophy, TPM can and does provide a way of breaking free from the traditional burden of production management. Managers who have made the transition will testify to the fact that managing a TPM factory, although still demanding, is substantially better than their previous situation.

Appendix 1 provides an audit checklist that can be used to assess the present position of the reader's operating company. It is suggested, at this point, that the reader takes a little time to complete the checklist for his/her company or area of the company before reading further. If appropriate, ask the opinion of colleagues who know the company/area well, but remember that each answer has to be yes or no and if there is any doubt, decide whether it is more yes than no or vice versa.

The checklist encourages the auditor to make an objective assessment of both the state of the production facilities used within the company or area of the company and also the prevailing attitudes. Many conclusions can be drawn from the general level of cleanliness and tidiness prevailing within an operating company and the checklist provides a diagnosis of the present status of the company based upon the number of 'yes' answers. I have used this checklist and variations on it on many occasions and to date have found that it is quite accurate in its conclusions.

■ 2.2 TPM and operational re-design

The operating companies of today have to respond to the ever-changing needs and demands of their business environment. Those that do not, or cannot because they are not equipped to respond, will be doomed to failure. Many companies have recognised this and have undergone a fundamental change in the way in which they operate and organise themselves.

The model illustrated in Figure 2.1 shows that an operating company carries out two main processes, namely:

1. The Development Process. The design and development of new products

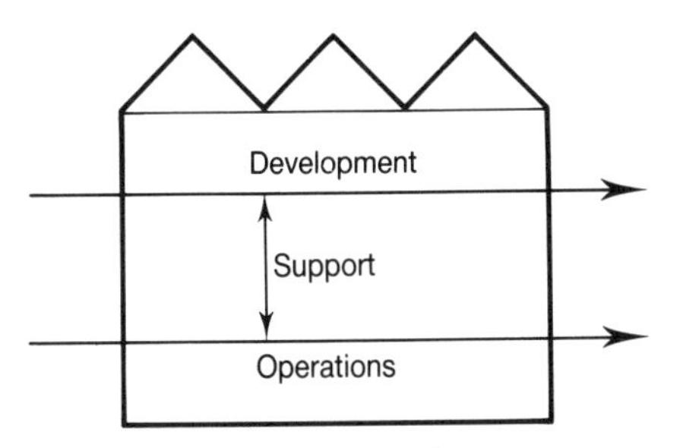

Figure 2.1 The two main processes

that the market will require along with the development of new processes and technology to manufacture the new products.

2. The Operations Process. The manufacture and sales of existing products using existing processes and technology.

All other functions within the business are there to support these two processes. Companies should, therefore, be organised in such a way that they support these two main processes and can carry them out most effectively. In practice, however, many businesses have become orientated around departments or functions, the growth and proliferation of which have been affected by company politics, senior management preferences and 'empire building'.

Operating companies are also affected by many external influences such as the following, illustrated in Figure 2.2:

- Market demand
- Customer taste and fashion
- The economy of particular nations
- Standards and new legal codes, etc.

A business does not usually have very much control, if any, over these external factors and, therefore, must be able to respond when any of them change. Within its own boundaries, however, a business does have control over the internal factors which govern how effectively it operates and its ability to respond to external changes. Figure 2.3 shows these internal factors.

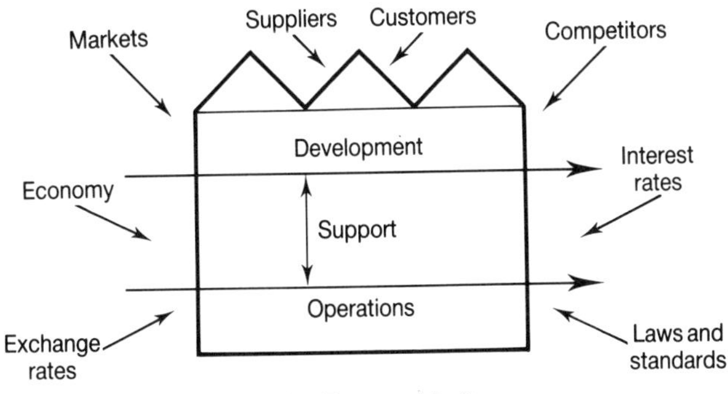

Figure 2.2 External influences

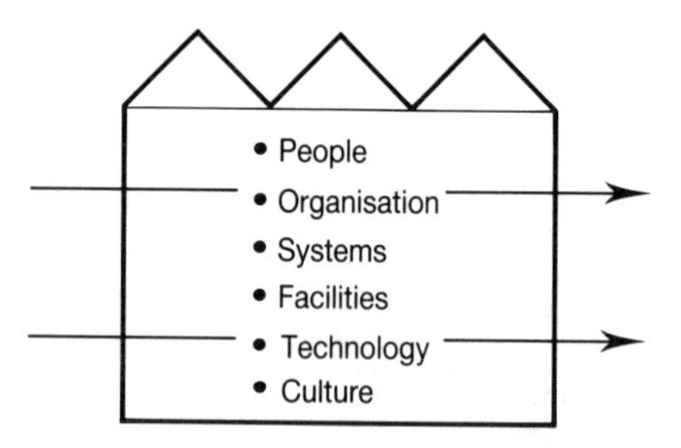

Figure 2.3 Internal factors

These internal factors can be described as follows:

- People. The skills and expertise of the people employed by the business, their enthusiasm, flexibility and how effectively they are utilised.
- Organisation. The way in which the company is organised, how effectively it achieves the main business processes, how quickly it can respond to external changes and how well it communicates.
- Systems. The systems which support the main business processes, how cost-effective they are, how complex or difficult to operate and how the different systems integrate within the company and interface with external systems.
- Facilities. The facilities used to carry out the main business processes, how effective they are, how flexible and responsive to varying customer demands they are, how costly they are to maintain and the financial return that they achieve.
- Technology. The type and level of technology that is employed within the company, how up to date and effective it is, how well it is supported and understood and the competitive advantage that it provides for the business.
- Culture. The prevailing culture within the company, people's attitudes and values, their relationships with colleagues and external parties. This is formed over many years and is rooted in the history of the business.

When a company is faced with a substantial gap between its own performance and that of its nearest competitors as depicted in Figure 2.4, it is necessary to take some radical action.

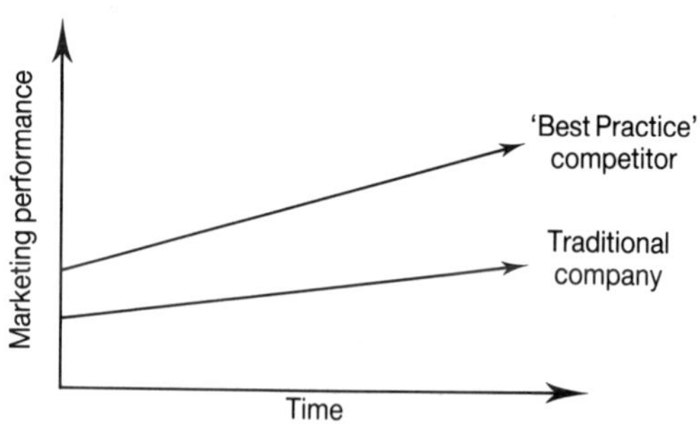

Figure 2.4 The performance gap

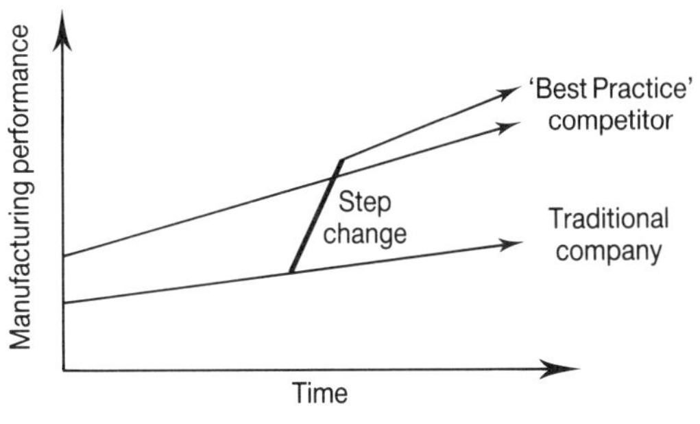

Figure 2.5 A step change

Small, incremental improvements which can be achieved through improvement group activity are just not enough in these circumstances and a step change in performance, such as illustrated in Figure 2.5, is needed.

An operational strategy has to be formulated which recognises the present position and what is required in order for the company to become competitive and profitable. The identification of the major building blocks of world-class manufacturing performance and how they can be addressed is a key part of the strategy. Implementation of the strategy can only be achieved through a number of projects aimed at changing the organisation, systems and methods of operation of the business.

Figure 2.6 shows these main areas or 'building blocks' for a typical operating company which can be described as follows:

Strategic direction:

- a high level strategy for the future of the business clearly communicated to all personnel.

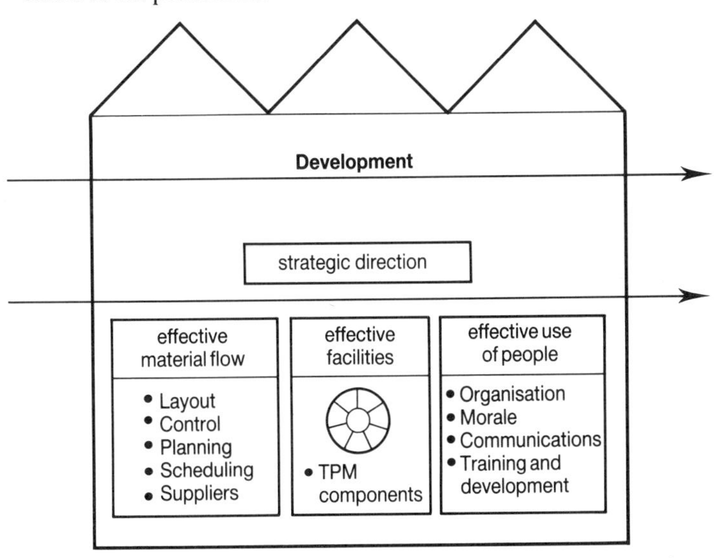

Figure 2.6 The building blocks of world-class manufacturing performance

Effective material flow:

- the layout and configuration of the factory floor to provide the smooth flow of materials through the factory and within each area or cell with minimum stock levels;
- simple, visual methods of the control of material flow, work in progress and parts identification;
- production planning and liaison with sales;
- long-, medium- and short-term scheduling of production including 'levelled scheduling' of resources;
- supplier rationalisation and development.

Effective facilities:

- the implementation of the components of Total Productive Maintenance (as described in Chapter 3).

Effective use of people:

- simple, flat organisational structure with minimum levels of supervision;
- a working environment which encourages good morale, enthusiasm and continuous improvement;
- open and honest, two-way communications which lead to mutual trust, teamwork and flexibility;
- training and development programmes for all personnel.

Many of these change projects are best undertaken by a multidisciplinary 'task force' team who are assigned the task of analysing the present situation and redesigning the business operations to achieve this step change in operating performance. Typically, this will mean physically reorganising the manufacturing area, installing new control systems, changing job roles and organisational structure and introducing new working practices.

The new design should incorporate 'best practice' features such as simple material flows, natural groups of processes and people and minimal waste. What should not be overlooked when planning and implementing a new manufacturing area is the two main resources required to make the new design work, namely people and machinery. Many operational redesign projects have failed, despite the valiant efforts of the task force team members, because these important aspects have been given too little effort, too late.

TPM is an essential part of a 'best practice' operational strategy and as such should be at the very heart of the way in which the new manufacturing area will operate. It is acceptable to launch TPM once the new layout and systems have been implemented, provided that there has been the appropriate involvement of people in the redesign process. If the new design has been imposed with little communication between the task force and factory floor personnel, then the risk of resistance, lack of co-operation and even confrontation is greatly increased. In these circumstances, the introduction of TPM will be very slow and difficult to achieve.

A much better approach is to launch TPM in the area either prior to the task force launch or in parallel with the design and implementation of the new manufacturing area. This has the dual benefits as follows:

- It involves people in the improvement of existing facilities, in the specification and purchase of new facilities and provides an interface with the task force team.
- It ensures that the facilities which will be used in the new manufacturing area are in good condition and have been improved by the very people who are going to operate, set up and maintain them.

Suppliers are also an important part of the production team and have a vital role to play in providing reliable deliveries of good quality parts. It is not sufficient to demand that suppliers meet new quality standards and delivery requirements. They should be encouraged and helped to implement TPM as this will undoubtedly improve the supply of good quality parts and, therefore, the performance of the operating company.

2.3 TPM and Total Quality

Total Quality (TQ) or Total Quality Management (TQM) has been widely discussed over the past ten years and many operating companies have attempted to introduce and implement it in one form or another. Some applications have been very successful, others moderately so and some have failed miserably. Total Quality is a business-wide philosophy which is all about changing attitudes, working practices, values and the overall method of operation of the company. Its overall aim is to improve continuously the operating performance of the business, thus providing better customer service and increased profitability. I have visited many operating companies in the past ten years, many of whom have been running TQM initiatives, sometimes for five years or more. Unfortunately, with some, TQM is only 'skin deep', i.e. it has been a relatively high-level exercise involving glossy posters, flow charts and other documents, mainly undertaken as a public relations exercise for their customers.

The achievement of national, international or industrial accreditation to particular quality standards does not necessarily ensure quality, or that the organisation has embraced the principles of Total Quality. What it means is that the existing systems and procedures within the business have been documented and means of control have been put in place. This is irrespective of whether these systems and procedures are effective or not. Accreditation is often demanded by customers and may be a qualifying criterion for operating in particular markets, and in these circumstances it must be achieved. We should not, however, confuse accreditation with the achievement of 'best practice' within operating companies.

A tour around the factory floor and a few 'off the record' discussions with factory personnel very soon show that the true philosophy behind TQM has not been applied. There has been a tendency to ignore activities aimed at improving the condition and performance of facilities altogether, especially at factory floor level.
The way in which TQM has been implemented in many businesses has tended not to address the 'real' people issues and the resulting TQM business may look impressive from the outside but, underneath, many of the old attitudes, prejudices and values still remain.

Communications within the company is always an area which appears to need improvement and is usually part of a Total Quality initiative. The resulting 'improved' communications structure very often only supports one-way communications, i.e. the top-down communication of management information and intentions. It misses out what is the most important part of communicating, i.e. listening. Effective channels for bottom-up communications are so often neglected, but are equally as important as top-down mechanisms.

A point that should be considered regarding communications is that personnel at different levels of the business tend to have a different business interest span as shown in Figure 2.7. At the top of the 'pyramid' is the Managing Director or General Manager whose business interest span is usually measured in one to five years as a major part of his job is to plan ahead. The senior managers or executives of the business will also think in terms of years, probably one to three years, and their subordinates, the middle managers, will tend to be interested in months to a year. Line managers to supervisors will be concerned with weekly and monthly targets and do not often have the opportunity to think further ahead than this, whereas factory floor personnel have an interest span of days to a few weeks.

This business interest span pyramid should have a bearing on the way in which plans and initiatives are communicated at different levels. If, for instance, a new Total Quality initiative is launched, the effects of which will not be seen at the factory floor level for months or years, then it will quickly lose

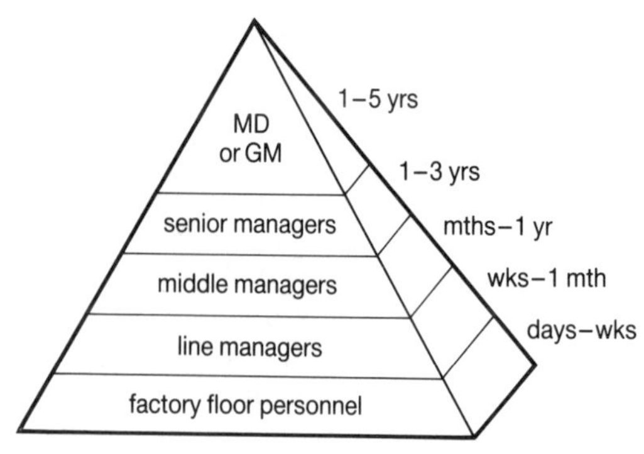

Figure 2.7 The interest span pyramid

credibility if it is communicated too early, i.e. more than weeks before anything tangible happens. By 'singing the praises' of some new innovation within the company to personnel who will see little or no effect in the short to medium term, the result is that they will only feel let down and the 'failure' will only fuel the perception that management is incompetent and untrustworthy. We will explore the issue of communications and the business interest span pyramid in more detail in Chapter 5.

TPM is Total Quality aimed at improving the condition and performance of the facilities that the business uses to perform the operations process. It is total quality at the 'sharp end', i.e. where processes are performed, value is added and the wealth is created for the business. Through the TPM teams, good and effective channels of communication are established and people at all levels within the company are given the opportunity to put forward their views and ideas. This is a very important issue for the business, and if Total Quality fails to address this area then it will not succeed. Therefore, we can see that TPM is an essential component of TQM. Figure 2.8 illustrates the scope of Total Quality and TPM within an operating business.

As part of a TQM programme many companies set up and run quality or process improvement teams which focus on particular machinery or process problems using a multidisciplinary team. The team will analyse the problem, brainstorm different options, decide upon the best solution and implement the improvement. This approach can be very effective and some excellent solutions to long-running problems have resulted. Such teams do not conflict with TPM in any way. As we will see in Chapter 3, one of the component parts of an overall TPM programme is the identification of the need for focused improvement projects and then setting up an appropriate team to implement them. Focused improvement projects fit neatly under the TPM 'umbrella', but TPM is definitely not a one-off project. It is a programme of change which will continue indefinitely.

TPM team activities will produce certain documents related to the maintenance, operation and performance of facilities and these can be used as

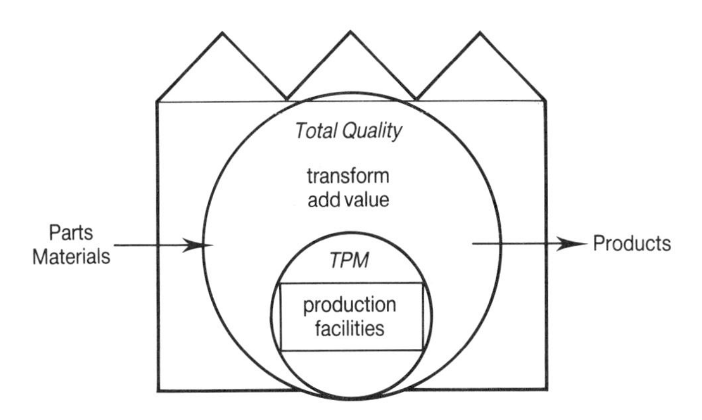

Figure 2.8 TPM and Total Quality

the part of the quality system documentation. Thus, TPM is complimentary to quality standards and systems and can be used to generate meaningful and effective procedures and work instructions.

■ 2.4 TPM and traditional maintenance practices

Reactive maintenance (i.e. breakdown) is all too common in our factories. It is a practice that is inherently wasteful and ineffective with disadvantages such as the following:

- No warning of failure. As no attempt is made to prevent or predict breakdowns then they usually happen with very little, if any, warning.
- Possible safety risk. When machinery parts fail there is a danger that they will fly off, shatter, cause loss of control, etc. thus causing injury to the operator. Also, many injuries occur whilst trying to investigate and rectify breakdowns.
- Unscheduled down-time of machinery. If failure is not predicted then breakdowns will cause the machinery to be unavailable for the production that will have been scheduled.
- Production loss or delay. If the machinery is not available then production schedules will be disrupted and production either delayed or stopped completely if no alternative machinery is available.
- Possible secondary damage. When machinery parts fail, they may cause loss of control which in turn leads to further damage. Often the secondary damage caused is much more significant and costly to repair than the original failure.

Breakdowns become accepted as a way of life and this leads to the need for:

- Stand-by machinery. Thus production can be loaded on to alternative machinery if a breakdown should occur.
- A stand-by maintenance team. Maintenance resources whose main function is to react to breakdowns when they occur.
- A stock of spare parts. In order to rectify breakdowns as quickly as possible there is a need to keep a substantial stock of spare parts.

There is a mistaken perception that breakdown maintenance is the cheapest approach to the maintenance of machinery that can be employed. This may be the case for some machinery, in particular that which is used infrequently, is not very complex and is relatively quick and easy to repair. It can be argued that even in this case, prevention of breakdowns is better than cure and simple, operator-based maintenance prevention activities could be employed which would reduce the number and frequency of breakdowns.

It is clear that in many cases it is desirable to move away from purely reactive maintenance towards more pro-active maintenance practices such as

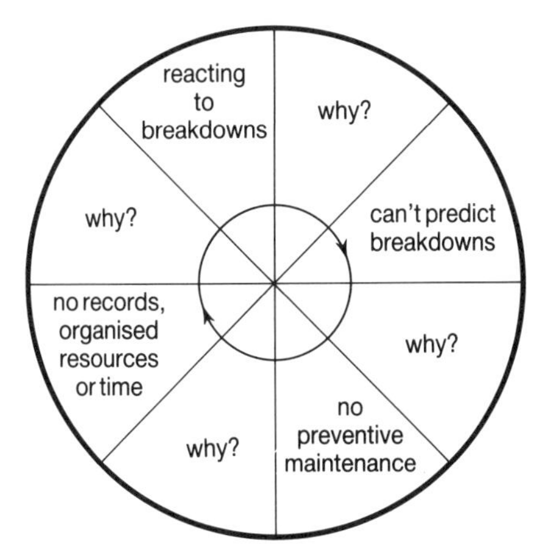

Figure 2.9 The vicious circle of reactive maintenance

planned, preventive maintenance and predictive maintenance. This can only be achieved if maintenance personnel are allowed the time required to plan, schedule and carry out preventive and predictive maintenance tasks and also to operate and update maintenance systems. Figure 2.9 illustrates the vicious circle of reactive maintenance that has to be broken.

If there is a combination of this vicious circle and no operator based, TPM-type activities, then machinery condition and the general state of facilities become locked into a downward spiral leading to further and further deterioration and more and more problems.

Figure 2.10 shows the downward spiral of machinery condition which leads to the following:

- An increase in the frequency and the number of breakdowns and also the time taken to fix breakdowns.
- A deterioration in the operating performance of machinery as its condition gradually worsens.
- An increase in the inconsistency of product output from machinery leading to more scrapped or re-worked parts.

This shows that this general neglect of facilities does not just increase breakdowns, but also performance and quality.

The real cost of reactive maintenance is much more than the cost of maintenance resources and spare parts as indicated in Figure 2.11. The costs include the following:

- Lost production: lost revenue from the sale of products, the cost of operators who cannot be deployed during the down-time, the loss of earning capacity.
- Disrupted schedules: panic measures to recover time lost such as premium time working and special delivery arrangements, disruption of other

Figure 2.10 The downward spiral of machinery condition

machinery schedules, loss of customer confidence leading to lost business and subcontract costs where outside help is needed.

- Repair costs: cost of maintenance staff to organise and carry out the repair, cost of any outside resources and services required.
- Stand-by machinery: the capital and running cost of any machinery kept for 'stand by', factory space costs.
- Spare parts: the cost of spare parts used (as a result of primary and secondary damage), storage costs for spare parts, premium costs to obtain parts that are not held in stock.

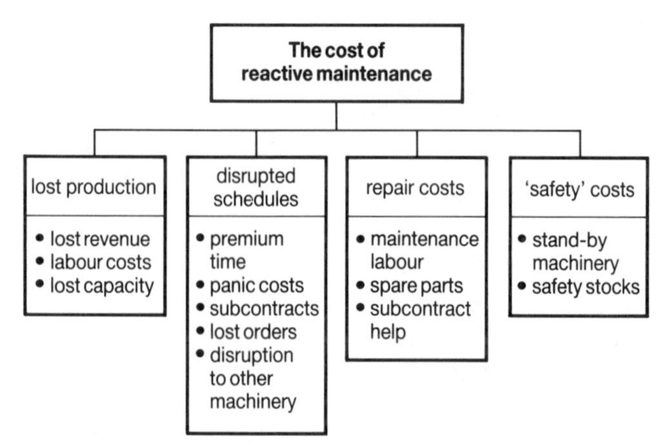

Figure 2.11 The real cost of reactive maintenance

- Additional work in progress: safety stocks held to cover for machinery unreliability, the cost of holding, storing and controlling stocks.

TPM has a definite role to play in reducing the amount of unskilled, relatively simple tasks that skilled maintenance professionals are expected to perform and also by improving machinery condition, which will have the effect of considerably reducing the amount of breakdowns. TPM will bring about a change in the traditional attitude of 'I operate the machinery, you fix it!' by promoting the production team and making the maintenance of machinery condition everyone's responsibility.

TPM is an approach which enables traditional maintenance practices to change from reactive to pro-active by sharing responsibility for machine condition, performance and maintenance. It enables the vicious circle of reactive maintenance to be broken.

2.5 TPM and maintenance systems

Machinery does break down from time to time and, although one of the aims of TPM is to eliminate breakdowns, it does happen. When it happens an effective system is needed for reporting, investigating, getting the repair done both quickly and cost-effectively and recording information about the breakdown.

Machinery operators and craftsmen have an important part to play in the system. TPM recognises this and will enable an effective breakdown maintenance system to be developed, operated and continuously improved. Also, TPM provides a number of mechanisms whereby breakdowns are analysed, the causes investigated and action taken to prevent further breakdowns.

Maintenance systems are also required for preventive and predictive maintenance tasks. An effective system is required to identify any checks that are needed, parts that may need replacement, adjustments that may need to be made and major overhauls or servicing work, the frequency, time required for them, etc.

By using TPM, preventive maintenance schedules can be made more meaningful as a result of the integrated approach of all production and maintenance personnel, and improvement in the information gathered and measures of performance used. Also, as discussed in section 2.4, TPM will 'free up' maintenance professionals and allow them the time required to carry out scheduled, preventive maintenance and, very importantly, will allow them to keep the system up to date so that preventive tasks, the frequency of carrying out these tasks, their relevance and their cost-effectiveness can all be reviewed continuously. This also applies to predictive maintenance schedules, particularly the development and application of predictive maintenance as an alternative to the preventive approach.

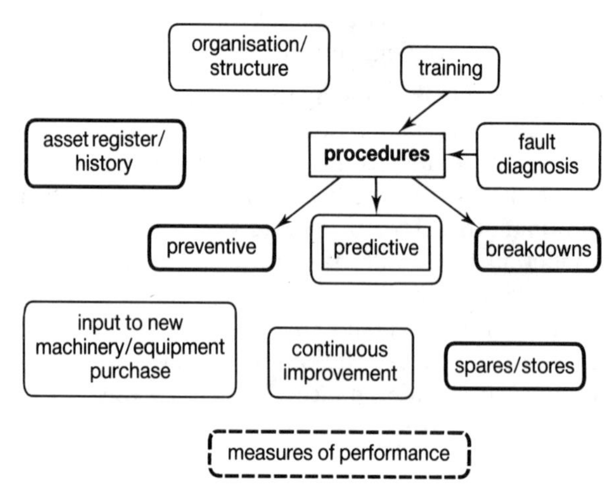

Figure 2.12 A maintenance system

Figure 2.12 shows the main elements of a maintenance system and its relationship with both TPM teams and business systems. Through TPM teams, information can be gathered concerning breakdowns, machinery faults, operating performance, overall effectiveness, etc. all of which provides an important input to the maintenance system. Also, as previously discussed, preventive maintenance schedules can be compiled accurately through the TPM teams. TPM will enable very effective maintenance systems to be developed and operated as an integral part of manufacturing operations.

■ 2.6 Maintaining the integrity of manufacturing operations

The 'maintenance' element of the title Total Productive Maintenance is misleading as it can give the initial impression that it is another maintenance technique. Unfortunately, as previously mentioned in Chapter 1, the traditional reaction will be to associate maintenance with spanners, dirty overalls and breakdowns, whereas TPM is trying to prevent breakdowns and all associated losses. This traditional Western interpretation of maintenance differs quite considerably from the Japanese one which appears to be much closer to 'maintaining the integrity of manufacturing operations', i.e. ensuring that manufacturing operations are carried out effectively, thoroughly and *properly*.

Maintenance activities are only part of the approach to maintaining the integrity of manufacturing operations and are inextricably linked with quality, performance, safety and morale. We need to work at changing perceptions and attitudes to maintenance and manufacturing operations within our businesses to make them more in line with those of our best competitors.

3 The Components of TPM

What does TPM consist of? TPM, in its widest sense, is both a philosophy and a collection of practices and techniques, all aimed at maximising the effectiveness of business facilities and processes.

The 'practical' components of TPM are contained within the TPM pie, as illustrated in Figure 3.1. The components can be given different titles, but are best described as follows:

- Understanding the status of the facilities that are being addressed by the TPM team, restoring them to a working condition, maintaining that condition and continuously improving the facilities. This is achieved through TPM teams consisting of people who operate, set up and maintain

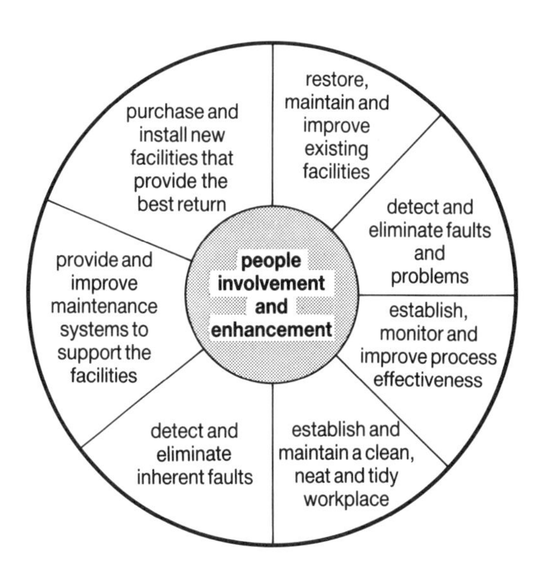

Figure 3.1 The practical components of TPM

the facilities as part of their normal working day and, thus, know the operating methods, problems and faults very well.

- Detect and eliminate faults and problems. Through the TPM team, faults which are present in the facilities are identified, their cause investigated and systematically eliminated. Also, operating difficulties which cause problems for the operator, setter or maintenance personnel and reduce the effectiveness of the process are eliminated.
- Establish, monitor and improve the effectiveness of the process that is carried out using the facilities. It is important that the TPM teams and the company understand just how well the process is being performed, and recognise the areas which are limiting or adversely affecting performance. This will enable effectiveness to be constantly reviewed and provide a focus for improvement activity.
- Establish and maintain a clean, neat and tidy workplace. The cleanliness of all facilities, desks, tables, benches, cupboards, cabinets, etc. along with the floor, walls and general environment is given a high degree of importance. All tools and materials used to achieve the process are identified, rubbish thrown away and the discipline of 'a place for everything and everything in its place' established.
- Detect and eliminate inherent faults. The activities of the TPM teams show up where facilities, processes or systems used have inherent faults which cannot be eliminated without a more fundamental re-design. Where these are detected, then focused project teams can be used to re-design and implement the changes.
- Provide and improve maintenance systems to support the facilities. More specialist skills will, in most cases, be needed to provide breakdown, servicing and improvement support for facilities, particularly those involving complex controls and mechanisms. Systems and resources for providing this support will need to be designed, implemented and continuously improved.
- Purchase and install new facilities that provide the best return. Any new purchases should incorporate any improvements and features that the TPM teams may have implemented on existing facilities. An excellent opportunity arises when purchasing new facilities to influence their design and operation so that greater levels of effectiveness can be achieved.

At the very heart of all of these TPM activities is the involvement and enhancement of people at all levels in the business, but most specifically the TPM team members.

It should be noted that all of the activities contained within the right-hand 'slices' of the TPM pie are carried out by production personnel in conjunction with engineering. This is a very important point and needs to be communicated clearly because the success of TPM does depend upon its philosophy being accepted by both production and engineering functions.

The components of TPM are relevant for many areas of application. In the case of a production area, the facilities consist of the machinery used to process materials: in assembly areas, the equipment, tools, etc. used to assemble are considered; in a laboratory environment, all of the equipment and instruments used to carry out tests and experiments are considered and in an office, the computers and office equipment are considered. In each different area the emphasis of TPM will vary as will the way in which TPM is applied, but the basic principles and philosophy are common to all environments.

TPM addresses all aspects of the condition of the facilities such as quality, operation, maintenance and improvement and provides a thorough, 'common sense' approach. The 'practical' components of TPM which are contained within the TPM pie will be explained in more detail in sections 3.2 to 3.8.

■ 3.1 TPM – an all-embracing philosophy

Before taking a more detailed look at the 'practical' components, it is important that the philosophy behind TPM is clearly understood. The 'practical' components are there to support the process of change which TPM brings about, and it is the change in attitudes, of values and the very culture of the business which is the aim of TPM, not the implementation of any particular technique or approach.

If TPM is to be applied successfully within any company there is a need to take on board not only the practices and techniques described but also the philosophy which is the very essence of TPM. It represents a fundamentally different style of management and employee participation. It is harder to resolve the 'soft', people issues than to apply the TPM practices and techniques and the people issues will, in most cases, take much more time and effort. The TPM philosophy contains the following elements:

- Team working.
- Respect for people at all levels.
- Motivation of people at all levels.
- Participation and encouragement.
- Positive leadership and support.
- Opportunity for people to acquire and enhance skills and experience, and develop their full potential.
- Continuous improvement, always striving to do better.
- Recognition of effort and providing incentives.

This is far removed from the situation which prevails in many businesses and this situation will have to change if TPM is to be implemented successfully and benefits are to be realised but, conversely, TPM can be used as a means of leading the change. Figure 3.2 illustrates the philosophy of TPM. The philosophy has to be accepted by all levels of the business if TPM is to really succeed. This will not be easy and may take many years to achieve.

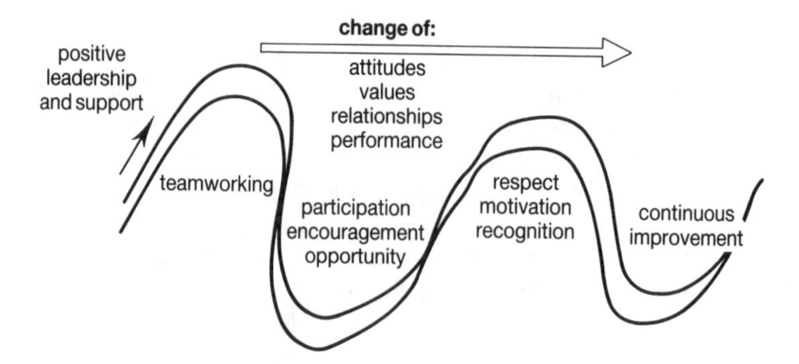

Figure 3.2 The TPM philosophy

3.2 Establish the condition of facilities, restore, maintain and improve them

Due to the typical situation which prevails and has prevailed for many years in operating companies (as discussed in Chapter 2), the general condition of facilities has deteriorated and this is especially true for production machinery and equipment. As part of my work with TPM teams I often encounter production machinery which is in a very poor state of repair and even fairly new machinery which has deteriorated in a matter of no more than one to two years. It would be ideal if no faults were present and this would certainly accelerate the TPM implementation, but the reality is that, in almost every case, faults and operating problems exist. It is not sufficient just to state that machinery is in a poor condition and needs to be restored. The details of all the faults and operating problems associated with the machinery are required and the best people to investigate these faults and operating problems are the people who operate, set up and regularly maintain the machinery in question.

Restoring machinery to a 'basic condition', i.e. eliminating all of the existing faults and problems is a prerequisite to maintaining this basic condition. What is meant by basic condition is that the machinery operates at a level of performance that is expected and that it was designed to achieve, and also that the operator is able to perform his/her function without too much difficulty. Once the machinery has been restored to a 'basic condition' then it is much more reasonable to detail the activities required to maintain that basic condition.

The TPM activities aimed at maintaining the basic condition of facilities are often referred to as 'autonomous maintenance' activities. Autonomous means self-governing, independent and devolved to a lower level, and in the context of TPM it means that certain simple, team-based activities are carried out by personnel in the locality of the particular machinery. Many Japanese texts when translated, use the word 'localisation' which can be interpreted to mean that TPM activities are based around the machinery location and are carried out by personnel who are local to the machine.

The people who operate the machinery must 'own' it and be encouraged to take pride in *their* machinery, take responsibility for its condition and seek to continually improve it. There may be a shared ownership across shifts or within an area or process where flexible working practices are employed. Through a structured TPM implementation process, local personnel are trained, developed and encouraged to clean, lubricate and inspect their machinery, check bolts, monitor its performance, draw up and use procedures and carry out relatively straightforward repairs.

The relative roles of operational and maintenance personnel are shown in Figure 3.3, but it is important that a production 'team' is built with maintenance personnel playing a key role. Localisation, as can be seen from Figure 3.3, means that the duties of the operational personnel include the following tasks:

- Major responsibility for the maintenance of basic machinery conditions to ensure that deterioration is prevented.
- Monitoring all aspects of machinery performance, measuring and detection of any deterioration.
- Carrying out regular inspection and detection and/or prediction of problems with the machinery.

The TPM manufacturer

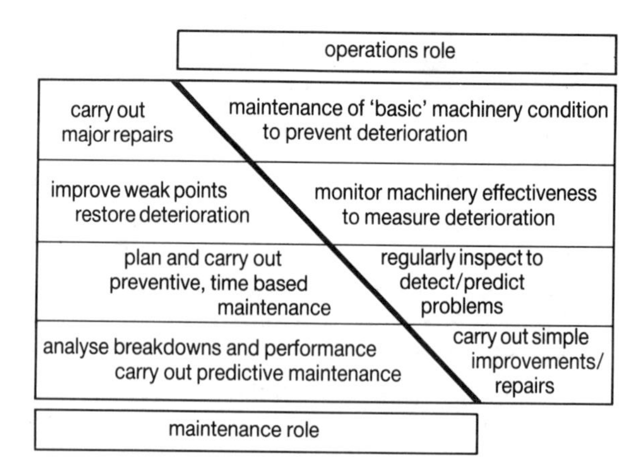

Figure 3.3 The roles of operations and maintenance personnel

- Carrying out simple repairs where reasonable to do so and also implementing small improvements.

They 'own' the machinery in their locality and share with maintenance personnel the responsibility for all aspects of the operation, condition and performance of the machinery. Localisation is a very practical concept which is central to TPM and promotes a fundamental change in attitude from operations, maintenance and management.

Having restored facilities and put in place activities to maintain a basic condition, along with practical mechanisms to support these activities, the natural progression is to look for ways of further improving the condition, performance and method of operation of the facilities. This is a continuous improvement approach which can only be effectively employed once the restoration has been undertaken and 'autonomous' maintenance activities have been established and institutionalised. Note that this component of TPM (which consists of three distinct stages) is production-driven and assisted by maintenance personnel.

3.3 Identify and eliminate faults and operating problems

What is the difference between faults and operating problems? Faults are things that are wrong with the machinery which either impair its operation now or are likely to do so in the future. Sometimes they are very minor faults (such as a broken door handle, compressed air leak or loose fastener) or more significant faults (such as worn slideways, defective bearings or leaking valves). Safety-related items are also included within this category (such as exposed wiring, inoperative guards or broken interlocks).

Operating problems do not necessarily impair the functionality of the machinery, but make it difficult to operate. These are usually things that make the normal operation of the machinery dirty (such as inadequate covers, oil leaks or the spraying of process media), dangerous (such as inadequate guarding, scattering of process media or product, slippery floors or inadequate access), or difficult (such as controls in the wrong place, inadequate lifting facilities, lack of foolproofing or the need to constantly re-adjust).

Figure 3.4 shows some of the types of faults and operating problems that are encountered such as the following:

Faults that:

- cause breakdowns and/or stoppages;
- will eventually cause breakdowns and/or stoppages;
- slow the process down;
- cause inconsistency within the process;
- cause sub-standard product to be manufactured;
- provide a safety hazard;
- will eventually cause a safety hazard;

Problems that:

- slow down the machinery operator;
- make changeovers and adjustments difficult and slow;
- make the operation of the machinery difficult;
- make the machinery and workplace dirty, oily and smelly;
- make the machinery dangerous to operate or set up;
- lead to injury or may eventually lead to strain injury.

This component of TPM is closely associated with establishing the condition of facilities and restoring them. It is by identifying faults and operating problems that the true and detailed condition of machinery is established and a plan of action compiled to rectify each fault and operating problem. Rectifying faults and operating problems may require various skills, materials and parts and these have to be identified and organised.

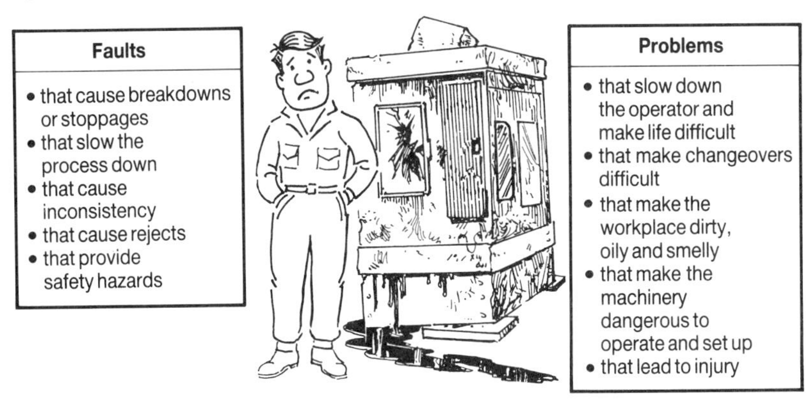

Figure 3.4 Types of faults and operating problems

■ 3.4 Measure the effectiveness of facilities

As previously stated, TPM is aimed at maximising the effectiveness of facilities and thus the process which they help to carry out. All facilities suffer from losses of some kind, i.e things that prevent them from operating effectively and which are caused by faults and operating problems. In order to improve the effectiveness of facilities we have to recognise, measure and reduce these losses. In TPM terms they are known as the six big losses and are described as follows:

1. Breakdown losses due to failures and repairs being carried out.
2. Set-up and adjustment losses which arise when changing from one product to another or one batch to another.
3. Idling and minor stoppage losses which arise due to sticking, faulty sensors, etc. and also waiting for material/parts to arrive or be loaded.
4. Speed losses where machinery is not operating at its optimum cycle time due to some fault or accumulation of faults.
5. Scrap and re-work losses arising from process capability problems, material problems, etc.
6. Start-up losses which can occur when a process takes some time or number of products to achieve 'steady state' or capable production.

The first two losses (breakdown and set-up/adjustment) can be grouped under the heading of 'availability', the next two (idling/minor stoppages and speed) under the heading of 'performance' and the last two (scrap/re-work and start up) under the heading of 'quality'. Let us look at some examples of the losses encountered in different areas of an operating company.

Example 1. High volume production

In this example we look at a special purpose drilling and tapping machine which makes a range of electrical terminal blocks. Working conditions are two shift working, 8 hours per shift, 5 days per week; planned throughput is 60 units per hour; actual output = 3320 units per week. The following is a list of losses encountered during the machining process:

1. The circular saw blade which cuts off the material shatters and has to be replaced. This happens once per week and takes 30 minutes.
   While the blade is being replaced the machine is not available for production, therefore this is an availability loss.
2. The saw pivot arm gets so congested with swarf and oil it becomes stiff and will not function properly. It has to be dismantled and cleaned. This happens twice per week and takes 45 minutes.
   While the pivot arm is being dismantled and cleaned the machine is not available for production, therefore there is an availability loss. Also, it is

highly likely that congestion of the pivot arm happens gradually, and a loss of performance and perhaps quality problems arise prior to the breakdown.

3. Cutting fluid is sprayed onto the bar feeder causing the bar to slip and not feed properly with two consequences:
   (a) if a part feed occurs then the block is cut off too short. This happens three times per day, takes 10 minutes to clear and 3 parts are lost.
   (b) if the bar does not feed then the machine stops, has to be cleared and re-set. This happens twice per week and takes 45 minutes.

   When (a) occurs then a minor stoppage happens which causes a performance loss, and also products are lost which causes a quality loss. When (b) occurs then more substantial down-time is encountered, causing lost production and, therefore, this is an availability loss.
4. On the first drilling head the main bearings get dry (no lubricant) causing the head to slow down and then seize up. This happens once every 6 weeks, the machine operates at half speed for one day prior to breaking down and to repair and replace the head takes 18 hours.

   When the drilling head seizes up, immediately the machine is not available for production and an availability loss is encountered. Just prior to the breakdown the drilling head operates at a speed which is approximately half of what it is designed to achieve and, therefore, a performance loss is suffered.
5. On the second tapping head the tap breaks and is undetected meaning that parts have to be tapped by hand. This happens once every day, 20 parts have to be re-worked, 10 parts are scrapped and it takes 15 minutes to replace the tap.

   Because the tap has broken it means that the machine is not achieving the complete process cycle and parts have to be re-processed, by hand, off the machine. This is the same as having to re-manufacture the 20 parts and thus this is a quality loss. Also, because some parts are damaged and cannot be re-processed they also represent a quality loss. The tap has to be replaced, and whilst the parts are cleared and the tap replaced, the machine is not available for production. Therefore, an availability loss is encountered (note that this could be interpreted as a performance loss, depending upon the criteria chosen).
6. To set up the machine for a new block, the machine parts have to be swapped and re-set and the stroke of the drills adjusted. This happens four times per week and takes 2½ hours.

   In this case the machine is not available for production while it is being set up for another product and, therefore, this is an availability loss.
7. Swarf builds up at the rear of the machine and has to be shovelled into a barrow. The operator has to stop the machine whilst doing this. This happens 3 times per day and takes 10 minutes each time.

   The swarf build-up causes a minor stoppage to occur when the operator has to stop the machine and clear it. This affects the performance of the machine and is, therefore, a performance loss.

8. The operator has to wait for the compressed air pressure to build up at the start of each shift. This takes 15 minutes each day.
   The operator is available but the machine cannot be operated for the first 15 minutes. In this case, therefore, an availability loss is incurred.

If we now add up all of the losses we find:

availability losses	=	30 mins × 1 (No. 1) + 45 mins × 2 (No. 2) + 45 mins × 2 (No. 3(b)) + 18 × 60/6 mins (No. 4 average) + 15 mins × 5 (No. 5) + 2½ × 60 mins × 4 (No. 6) + 15 mins × 5 (No. 8)
total	=	1140 minutes per week
performance losses	=	10 mins × 15 (No. 3(a)) + 4 × 60/6 mins (No. 4) + 10 mins × 15 (No. 7)
total	=	340 minutes per week
quality losses	=	3 parts × 3 × 5 (No. 3(a)) + 20 parts × 5 (No. 5) + 10 parts × 5 (No. 5)
total	=	195 parts per week

Example 2. Hand assembly

In this example we look at the assembly of machinery on the shop floor of a major machinery supplier's factory. Working conditions are single shift working, 8 hours per shift, 5 days per week; planned assembly loading time for this job is 8 weeks. The following is a list of losses encountered during a particular assembly process:

1. An alignment fixture was damaged and had to be fixed prior to use, 8 hours' delay.
   The fixture was required in order to carry on with the assembly process which could not happen because the fixture was not available. Therefore, this is an availability loss.
2. An important fixture was being used by another craftsman, on another assembly. This happened twice during the assembly, 3 hours' delay each time.
   Similarly, not in this case due to the fixture being damaged, but because it was being used elsewhere, the assembly could not continue as the fixture was not available and an availability loss was encountered.

3. The hand drill overheats and overloads. The craftsman had to 'peck drill' and leave the drill to cool off for 10 minutes. This happened on average once per day.
   The facility being used to perform the assembly operation, in this case a hand drill, was not performing correctly and causing minor stoppages. Therefore, this is a performance loss.
4. The craftsman had to go to the stores for the correct cutting tool. This happened once per day and took on average 12 minutes each time.
   Whilst visiting the stores, the craftsman could not work on the assembly. This happened because the correct tool was not available and, therefore, this is an availability loss. Note that this could have been classed as a minor stoppage, i.e. a performance loss depending upon the criteria applied.
5. The shop crane was not available as it was being used elsewhere. This happened on average once per day and caused a delay of 15 minutes each time.
   The shop crane was a key facility required to carry out the assembly process and was not available because it was being used by someone else. Therefore, this is an availability loss.
6. The assembly drawing was unclear, the craftsman needed to discuss the assembly with the designer. This happened on average twice per week and took 30 minutes.
   The drawing was an item of information supplied to the craftsman which he/she required in order to carry out the assembly process. As the drawing was not clear it can be interpreted that it was defective and thus time was lost due to the poor quality of information supplied. Therefore, a quality loss was incurred.
7. The correct spanner was missing and had to be retrieved. This happened twice per week and on average took 20 minutes each time.
   The correct spanner was required to carry out the assembly process and was not available. Therefore, this is an availability loss.
8. A part had been assembled, but due to incorrect dimensions had to be removed and re-machined. This happened 3 times during the assembly and each time took 5 hours.
   Time was lost because this part of the assembly had to be repeated due to the poor quality of parts. Therefore, a quality loss was encountered.
9. The test rig, used to carry out final testing of the machine, was not feeding parts correctly and had to be de-bugged and repaired. This happened once and took 6 hours.
   The test rig was a facility required to complete the assembly process and due to a fault it was not available. This is an availability loss.
10. The test rig had to be positioned and a transition chute made and fitted onto the assembly. This had to be done once and took 5 hours.
   The test rig had to be tooled, positioned and set up prior to the assembly process being able to continue and, therefore, an availability loss was encountered.

11. A part was fitted back to front by mistake and had to be removed and re-assembled. This happened twice during the assembly and cost 4 hours each time.
Time was lost and this part of the process had to be repeated due to the error of the craftsman involved. It may be that the part was not identified sufficiently or that the drawing was unclear and, therefore, the loss was due to poor quality and incurred a quality loss.
12. An important part was not available for assembly and no other work on the assembly could be done. This happened once and cost 8 hours.
Assembly operations are always very much affected by the shortage of parts and in this case an important part was not available for the assembly. Therefore, an availability loss was incurred.
13. Parts were not available when required to complete a sub-assembly and so that had to be left and another sub-assembly worked on. This happened 10 times and caused a delay of 15 minutes in changing over from job to job.
On these occasions the assembly process could continue, but the shortage of parts required caused disruption and minor stoppages resulted. This is a performance loss.

If we now add up all of the losses we find:

availability losses = 8×60 mins (No. 1) +
3×60 mins $\times 2$ (No. 2) +
12 mins $\times 5 \times 8$ (No. 4) +
15 mins $\times 5 \times 8$ (No. 5) +
20 mins $\times 2 \times 8$ (No. 7) +
6×60 mins (No. 9) +
5×60 mins (No. 10) +
8×60 mins (No. 12)

total = 3380 minutes during the assembly

performance losses = 10 mins $\times 5 \times 8$ (No. 3) +
15 mins $\times 10$ (No. 13)

total = 550 minutes during the assembly

quality losses = 30 mins $\times 2 \times 8$ (No. 6) +
5×60 mins $\times 3$ (No. 8) +
4×60 mins $\times 2$ (No. 11)

total = 1860 minutes during the assembly

Example 3. Semi-automated assembly

In this example we look at a semi-automated assembly machine which assembles and welds automotive components. Working conditions are single shift working, 7½ hours per shift, 5 days per week; planned throughput is 150 units/hour; actual output = 2875 units per week. The following is a list of losses encountered during the assembly process:

1. Two parts are not assembled correctly and do not sit properly in the fixture causing the machine to stop and they have to be taken out of the fixture and the machine re-set. This happens on average 5 times per hour, 1 unit is lost and 2 minutes is lost.
   Because parts have to be retrieved and the machine re-set, a minor stoppage occurs and, therefore, a performance loss is encountered. Also, because parts are incorrectly assembled and cannot be re-processed they are scrapped. This is also a quality loss.
2. Electrodes wear out and have to be replaced and the current level re-adjusted. This happens once per week, 30 units are scrapped and replacement takes 1 hour.
   While the electrodes are replaced and set up the machine is not available for production and, therefore, an availability loss is incurred. Also, as a result of the electrodes, wear not being detected early enough, a number of units are incorrectly welded and have to be rejected. Thus a quality loss is also incurred.
3. One of the electrode cooling hoses bursts, has to be replaced and the whole machine dried out. This happens once per month and takes 5 hours.
   Because the hose burst causes a breakdown during which time the machine cannot be operated, an availability loss is incurred.
4. One of the fixtures is slightly misaligned and can sometimes cause one of the welds to be out of position, 220 units per week are lost.
   The misaligned fixture causes inconsistent quality of products resulting in the rejection of a number of units. Therefore, this is a quality loss.
5. Fixtures have to be removed and replaced and electrode positions adjusted when changing to different size parts. This happens 3 times per week and takes 2.25 hours, 24 units are scrapped each time.
   Production time is lost whilst the machine is changed over and set up for a different product and, therefore, it is not available and this is an availability loss. Also, at start-up, units are lost and this represents a quality loss.
6. Sometimes the stamp (on the final station) actuating cylinder sticks and causes a delay which doubles the cycle time. This happens 30 minutes per day.
   When the cylinder sticks it reduces the output of the machine and, therefore, a performance loss is encountered.
7. In order to stop rusting of the machine parts (due to water leaks) the operator has to apply a protective spray at the start and end of each day. This takes 5 minutes each time.
   Because the operator cannot start production until he or she has applied the spray, a minor stoppage occurs which is a performance loss.
8. Limit switches on the indexing table sometimes corrode and stop the

machine from operating. They have to be replaced. This happens once every 6 weeks and takes 6 hours.
While the switches are being replaced, the machine is not available for production and, therefore, an availability loss is incurred. It is also likely that the switches could have caused some minor stoppages prior to breaking down which would represent a performance loss.

If we now add up all of the losses we find:

availability losses = 1×60 mins (No. 2) +
$5 \times 60/4$ mins (Av. No. 3) +
$2.25 \times 3 \times 60$ mins (No. 5) +
$6 \times 60/6$ mins (Av. No. 8)

total = 600 minutes per week

performance losses = 2 mins $\times 5 \times 7.5 \times 5$ (No. 1) +
15 mins $\times$ 5 (No. 6) +
5 mins $\times 2 \times 5$ (No. 7)

total = 500 minutes per week

quality losses = 1 unit $\times 5 \times 7.5 \times 5$ (No. 1) +
30 units (No. 2) +
220 units (No. 4) +
24 units $\times$ 3 (No. 5)

total = 510 units per week

Example 4. Processing system

In this example we take a food processing system which consists of a mixing vessel, hoppers holding raw materials, water tank valves and interconnecting pipework. Working conditions are three shift working, 24 hours a day, 7 days per week; throughput capacity = 300 kg/hour (batches of 75 kg); actual output = 43 000 kg. The following is a list of losses encountered during production:

1. A blockage occurs in the output pipe from one of the raw material hoppers. The blending process is delayed whilst the cover is taken off and blockage cleared. This happens on average 10 times per week and takes 8 minutes to clear.
   Whilst the blockage is being cleared, the process is delayed and a minor stoppage occurs. This is a performance loss.
2. Blended materials leak from a flange in the main output pipe from the mixing vessel. Around 3 kg of material is lost every shift and has to be cleared from the floor.
   The material which leaks from the defective flange joint has to be disposed of and, therefore, represents a quality loss.

3. The water inlet valve seizes and will not open. It has to be stripped down, cleaned and then re-fitted. This takes 4 hours to complete during which time the process cannot continue. Also, the material already in the mixing vessel has to be removed and disposed of. This happens on average once per week and 50 kg of material is wasted.
   The process cannot be performed whilst the repair is taking place as the system is not available and, therefore, this is an availability loss. Also, the waste material represents a quality loss.
4. The mixer paddle switch is faulty and often the mixing sequence stops until the switch is nudged by the operator. This happens around 12 times every shift and delays the processing of the batch by 3 minutes each time.
   The paddle switch fault causes a minor stoppage to occur during the process cycle. Therefore, a performance loss has been incurred.
5. A seal on the mixing vessel outlet ruptures and blended material spills onto the floor. The seal has to be replaced which takes 8 hours and approximately half of the batch is wasted and has to be disposed of. This happens once every 2 weeks.
   The seal rupture represents a breakdown and has to be repaired before the process can recommence. This is an availability loss. Also the wasted material means that a quality loss has been incurred.
6. One of the hopper feeders malfunctions and causes too much material to be added to the mix. Some of the mix has to be removed from the vessel and the remaining mix re-processed to attain the correct consistency. The mix that has been removed is later put back into the mixing vessel and re-processed. This happens to 4 batches per week and it takes twice as long to re-process the batch as usual.
   The amount of material that is re-processed is treated as if it were waste because the process has been undertaken twice and, therefore, a quality loss is incurred. Also, because the re-processing takes twice as long as a normal batch, it represents a reduced speed loss which is a performance loss.
7. Occasionally the main mixer drive overheats and causes the thermal overload to trip out. This requires a pause of 10 minutes before recommencing the process and occurs on average 6 times per week.
   The motor overload causes a minor stoppage which is a performance loss.

If we now add up all of the losses we find:

availability losses = 4 × 60 mins (No. 3) +
8 × 60 mins/2 (No. 5)

total = 480 minutes per week

performance losses = 8 mins × 10 (No. 1) +
3 mins × 12 × 3 × 7 (No. 4) +
15 mins × 4 (No. 6) +
10 mins × 6 (No. 7)

total = 956 minutes per week

$$\text{quality losses} = 3\text{ kg} \times 3 \times 7\text{ (No. 2)} + 50\text{ kg (No. 3)} + 37.5\text{ kg/2 (No. 5)} + 75\text{ kg} \times 4\text{ (No. 6)}$$

$$\text{total} = 431.75\text{ kg per week}$$

The reader should note that it is quite usual for different people to have different interpretations of the kind of loss which has been incurred. In particular, there is often a debate concerning what should be regarded as availability losses and what should be regarded as performance losses. I tend to think of minor stoppages as frequently occurring events which result in a production stoppage of 10 minutes or less, although this is not a hard and fast rule. It is necessary, therefore, to devise some criteria for each area where losses are to be measured so that everyone can record them under the same category and confusion is eliminated. The important principle is that all losses should be identified, and this is much more important than being able to place them in exactly the right category.

Similarly, types of losses can be identified which are associated with the facilities used to carry out a process in all areas of the business. Figure 3.5 shows the effects of the six big losses and how they reduce the productivity and, hence, the earning capacity of facilities. I have called the diagram an 'effectogram' as it provides a means of showing the six big losses and overall effectiveness using a histogram format. It is now an appropriate point to introduce the concept of overall effectiveness.

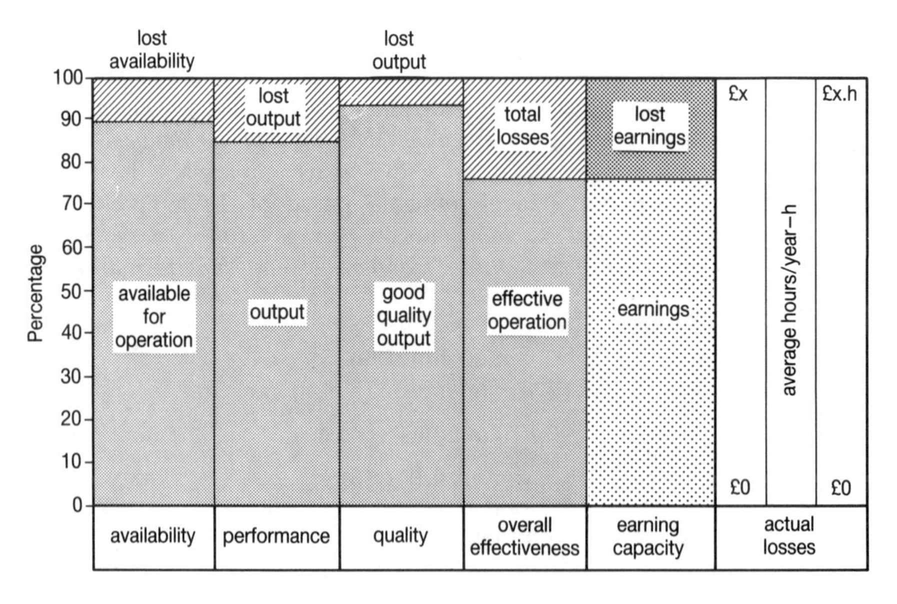

Figure 3.5 Effectogram showing the effects of the six big losses

We have, in the past, been encouraged to strive for efficiency and efficient manufacturing operations to produce goods at the highest volume for the least cost. Efficiency can be defined as 'doing things right', whereas effectiveness is 'doing the right things right'. In the context of an operating company this means producing what the customer wants, when he/she wants it, in the quantity he/she desires and at the appropriate quality and providing the required profit to the business: in essence, working in a smarter way.

Overall effectiveness

The effectiveness of facilities has a direct bearing upon the competitiveness and profitability of a business and maximising their effectiveness means that the best possible return is generated by each capital asset owned by the business. It is possible to calculate a percentage figure for each group of losses thus – percentage availability is the ratio of how long you actually used the machinery over how long you wanted to use the machinery, and is calculated as:

$$\%\ \text{availability} = \frac{\text{loading time} - \text{breakdown and set up time loss}}{\text{loading time}} \times 100$$

where loading time is the time that the machinery was planned to be in operation. A simple example is where loading time = 8 hours, breakdowns = 1 hour and changeovers/set ups = 1 hour, thus:

$$\%\ \text{availability} = \frac{8 - (1 + 1)}{8} \times 100 = \frac{6}{8} \times 100 = 75\%$$

There is often some debate concerning the definition of loading time and whether certain factors should or should not be included. It is important to remember that we are not trying to measure machinery or operator utilisation, but rather the availability of the machinery for production when it is required. My experience is that if people factors are included in the calculation then it is perceived as a measure of performance on the operator and not just the machinery, and as such can cause some resistance. Loading time may only be for a few hours a week, but if the machinery is required to be available for those few hours and is scheduled as such, then its percentage availability is based upon those few hours. My preferred definition of loading time is:

$$\text{loading time} = \text{planned production time} - \text{breaks} - \text{planned maintenance time}$$

Percentage performance is the ratio of what was actually produced in a given time over what you would have expected to be produced in a given time, and can be calculated in two ways. The first method is:

$$\%\ \text{performance} = \frac{\text{quantity produced}}{\text{time run} \times \text{capacity/given time}} \times 100$$

A simple example is where the quantity produced = 500 parts; the time run = 6 hours and the capacity = 100 parts per hour.

$$\% \text{ performance} = \frac{500}{6 \times 100} \times 100 = \frac{500}{600} \times 100 = 83\%$$

This is the most straightforward means of calculating percentage performance and is preferable where many products or bulk quantities are produced in a relatively short time. There are, however, situations where only few parts are produced per day or per week or even per month or year. In these cases standard production times are rarely used or accurate enough and, therefore, it is necessary to measure minor stoppages and reduced speed losses directly. In this case:

$$\% \text{ performance} = \frac{\text{time run} - \text{minor stoppages} - \text{reduced speed}}{\text{time run}} \times 100$$

A simple example is where time run = 6 hours; minor stoppages total = ½ hour lost and the reduced speed equivalent = ½ hour lost.

$$\% \text{ performance} = \frac{6 - \frac{1}{2} - \frac{1}{2}}{6} \times 100 = \frac{5}{6} \times 100 = 83\%$$

Note that the percentage performance figure can be calculated in either way, but it is usually simpler to use the first formula when reasonable quantities and standard throughput rates are available.

Percentage quality is the ratio of the number of good products over total products produced during a given period of time and, again, can be calculated in two ways. The first method is:

$$\% \text{ quality} = \frac{\text{amount produced} - \text{amount defects} - \text{amount re-processed}}{\text{amount produced}} \times 100$$

A simple example is where the quantity produced = 500 products; the amount defective = 50 and the amount re-processed = 50.

$$\% \text{ quality} = \frac{500 - 50 - 50}{500} \times 100 = \frac{400}{500} \times 100 = 80\%$$

As with the percentage performance calculation, this is the most straightforward way of calculating percentage quality where many products or bulk quantities are produced. Where this is not the case, it may be necessary to record the amount of time spent producing reject parts or work and the amount of time spent re-processing parts. In this case the calculation is:

$$\% \text{ quality} = \frac{\text{time run} - \text{defect time} - \text{re-processing time}}{\text{time run}} \times 100$$

A simple example is where the time run = 6 hours; the time spent producing defects = ½ hour and the time spent re-processing = ½ hour.

$$\% \text{ quality} = \frac{6 - \frac{1}{2} - \frac{1}{2}}{6} \times 100 = \frac{5}{6} \times 100 = 83\%$$

Note that the percentage quality figure can be calculated in either way, but it is usually much easier to use the first formula when the situation allows.

Overall effectiveness is a measure of all three of these factors and, although it is not strictly a percentage, it is usually represented in percentage terms and is calculated as:

$$\text{overall effectiveness} = \% \text{ availability} \times \% \text{ performance} \times \% \text{ quality}$$

In order to finish up with a percentage figure it is necessary to divide each individual percentage figure by 100 and then multiply the resulting overall effectiveness figure by 100. A simple example is:

% availability = 75%
% performance = 83%
% quality = 80%

$$\text{overall effectiveness} = 0.75 \times 0.83 \times 0.8 \times 100 = 50\%$$

If we now take the examples used to illustrate the six big losses, we can calculate an overall effectiveness figure for each example.

Example 1. High volume production

$$\text{loading time} = 8 \times 2 \times 5 = 80 \text{ hours/week} = 4800 \text{ mins/wk}$$

$$\begin{aligned}\text{availability losses} &= (30 \times 1) + (45 \times 2) + (45 \times 2) + (18/60/6) + (15 \times 5) \\ &\quad + (2.5 \times 60 \times 4) + (15 \times 5) \\ &= 30 + 90 + 90 + 180 + 75 + 600 + 75 = 1140 \text{ mins/wk}\end{aligned}$$

$$\% \text{ availability} = \frac{4800 - 1140}{4800} \times 100 = \frac{3660}{4800} \times 100 = 76\%$$

$$\begin{aligned}\text{performance losses} &= (10 \times 15) + (4 \times 60/6) + (10 \times 15) \\ &= 150 + 40 + 150 = 340 \text{ mins/wk}\end{aligned}$$

$$\% \text{ performance} = \frac{3660 - 340}{3660} \times 100 = \frac{3320}{3660} \times 100 = 91\%$$

OR

$$\% \text{ performance} = \frac{3320}{3660/60 \times 60)} \times 100 = \frac{3320}{3660} \times 100 = 91\%$$

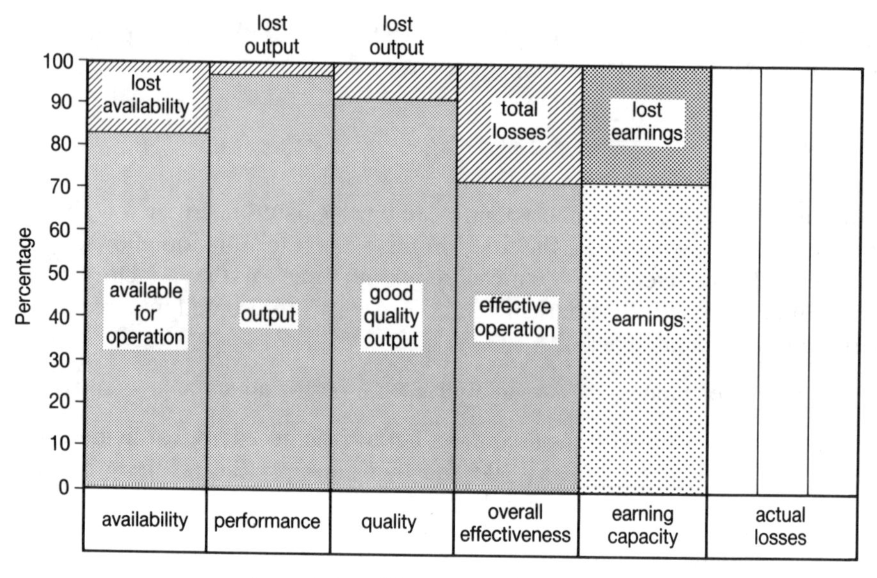

Figure 3.6 Effectogram for example 1

$$\text{quality losses} = (3 \times 3 \times 5) + (20 \times 5) + (10 \times 5) = 45 + 100 + 50 = 195 \text{ parts/wk}$$

$$\%\ \text{quality} = \frac{3320 - 195}{3320} \times 100 = \frac{3125}{3320} \times 100 = 94\%$$

$$\text{overall machine effectiveness} = 0.76 \times 0.91 \times 0.94 \times 100$$

OME = 65%

Example 2. Hand assembly

$$\text{loading time} = 8 \times 5 \times 8 = 320 \text{ hours} = 19\,200 \text{ mins}$$

$$\begin{aligned}\text{availability losses} &= (8 \times 60) + (3 \times 60 \times 2) + (12 \times 5 \times 8) + (15 \times 5 \times 8) + (20 \times 2 \times 8) + (6 \times 60) + (5 \times 60) + (8 \times 60)\\ &= 480 + 360 + 480 + 600 + 320 + 360 + 300 + 480\\ &= 3380 \text{ mins/assembly}\end{aligned}$$

$$\%\ \text{availability} = \frac{19\,200 - 3380}{19\,200} \times 100 = \frac{15\,820}{19\,200} \times 100 = 82\%$$

$$\text{performance losses} = (10 \times 5 \times 8) + (15 \times 10) = 400 + 150 = 550 \text{ mins/assembly}$$

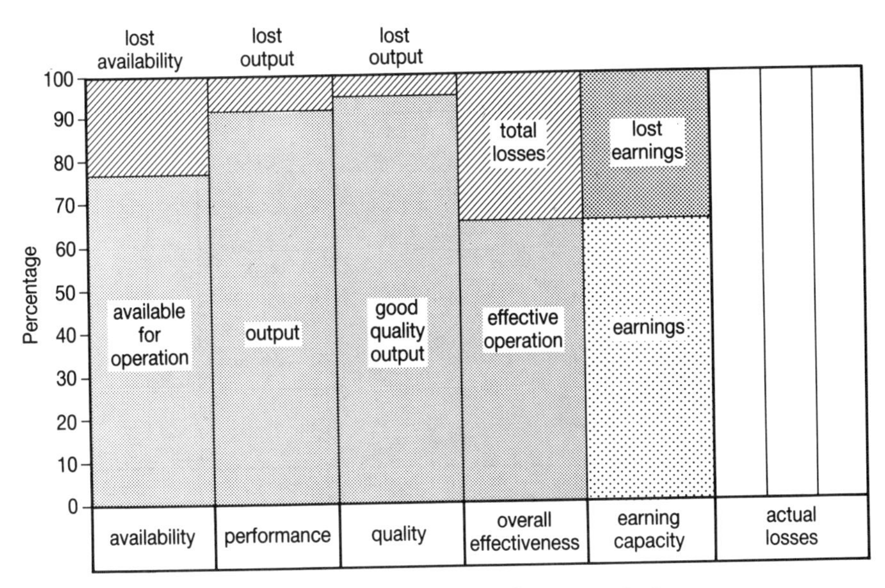

Figure 3.7 Effectogram for example 2

$$\% \text{ performance} = \frac{15\,820 - 550}{15\,820} \times 100 = \frac{15\,270}{15\,820} \times 100 = 97\%$$

$$\begin{aligned}\text{quality losses} &= (30 \times 2 \times 8) + (5 \times 60 \times 3) + (4 \times 60 \times 2)\\ &= 480 + 900 + 480 = 1860 \text{ mins/assembly}\end{aligned}$$

$$\% \text{ quality} = \frac{15\,820 - 1860}{15\,820} \times 100 = \frac{13\,960}{15\,820} \times 100 = 88\%$$

$$\text{overall assembly effectiveness} = 0.82 \times 0.97 \times 0.88 \times 100$$

$$\textbf{OAE} = \mathbf{70\%}$$

If we now plot these figures on an ‘effectogram’, the magnitude of the losses and loss of earning capacity can be easily seen. The effectogram for example 2 is shown in Figure 3.7.

Example 3. Semi-automated assembly

$$\text{loading time} = 7.5 \times 5 = 37.5 \text{ hours/wk} = 2250 \text{ mins/wk}$$

$$\begin{aligned}\text{availability losses} &= (1 \times 60) + (5 \times 60/4) + (2.25 \times 3 \times 60) + (6 \times 60/6)\\ &= 60 + 75 + 405 + 60 = 600 \text{ mins/wk}\end{aligned}$$

$$\% \text{ availability} = \frac{2250 - 600}{2250} \times 100 = \frac{1650}{2250} \times 100 = 73\%$$

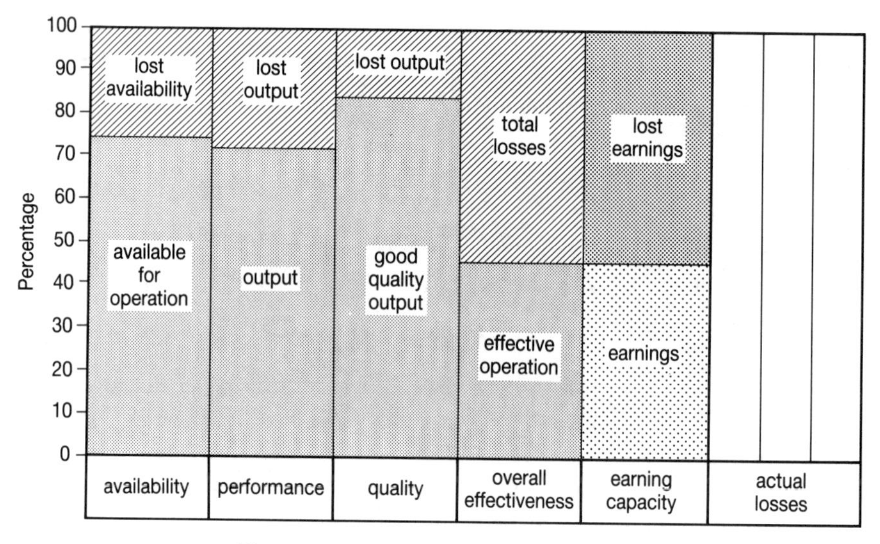

Figure 3.8 Effectogram for example 3

$$\text{performance losses} = (2 \times 5 \times 7.5 \times 5) + (15 \times 5) + (5 \times 2 \times 5)$$
$$= 375 + 75 + 50 = 500 \text{ mins/wk}$$

$$\% \text{ performance} = \frac{1650 - 500}{1650} \times 100 = \frac{1150}{1650} \times 100 = 70\%$$

OR

$$\% \text{ performance} = \frac{2875}{(1650/60 \times 150)} \times 100 = \frac{2875}{4125} \times 100 = 70\%$$

$$\text{quality losses} = (1 \times 5 \times 7.5 \times 5) + (30 \times 1) + (220 \times 1) + (24 \times 3)$$
$$= 188 + 30 + 220 + 72 = 510 \text{ units/wk}$$

$$\% \text{ quality} = \frac{2875 - 510}{2875} \times 100 = \frac{2365}{2875} \times 100 = 82\%$$

$$\text{overall machine effectiveness} = 0.73 \times 0.70 \times 0.82 \times 100$$

OME = 42%

The effectogram for this example is shown in Figure 3.8

Example 4. Processing system

$$\text{loading time} = 24 \times 7 = 168 \text{ hours/wk} = 10\,080 \text{ mins/wk}$$

$$\text{availability losses} = (4 \times 60) + (8 \times 60/2) = 240 + 240 = 480 \text{ mins/wk}$$

$$\% \text{ availability} = \frac{10\,080 - 480}{10\,080} \times 100 = \frac{9600}{10\,080} \times 100 = 95\%$$

$$\begin{aligned}\text{performance losses} &= (8 \times 10) + (3 \times 12 \times 3 \times 7) + (15 \times 4) + (10 \times 6) \\ &= 80 + 756 + 60 + 60 = 956 \text{ mins/wk}\end{aligned}$$

$$\% \text{ performance} = \frac{9600 - 956}{9600} \times 100 = \frac{8644}{9600} \times 100 = 90\%$$

OR

$$\% \text{ performance} = \frac{43\,000}{(9600/60 \times 300)} \times 100 = \frac{43\,000}{48\,000} \times 100 = 90\%$$

$$\begin{aligned}\text{quality losses} &= (3 \times 3 \times 7) + (50 \times 1) + (37.5/2) + (75 \times 4) \\ &= 63 + 50 + 18.75 + 300 = 431.75 \text{ kg/wk}\end{aligned}$$

$$\% \text{ quality} = \frac{43\,000 - 413.75}{43\,000} \times 100 = \frac{42\,586.25}{43\,000} \times 100 = 99\%$$

$$\text{overall plant effectiveness} = 0.95 \times 0.90 \times 0.99 \times 100$$

$$\textbf{OPE} = \textbf{85\%}$$

The figures are plotted on the effectogram in Figure 3.9.

The reader should note that the percentage figures have been rounded up to the nearest whole number in all of the examples.

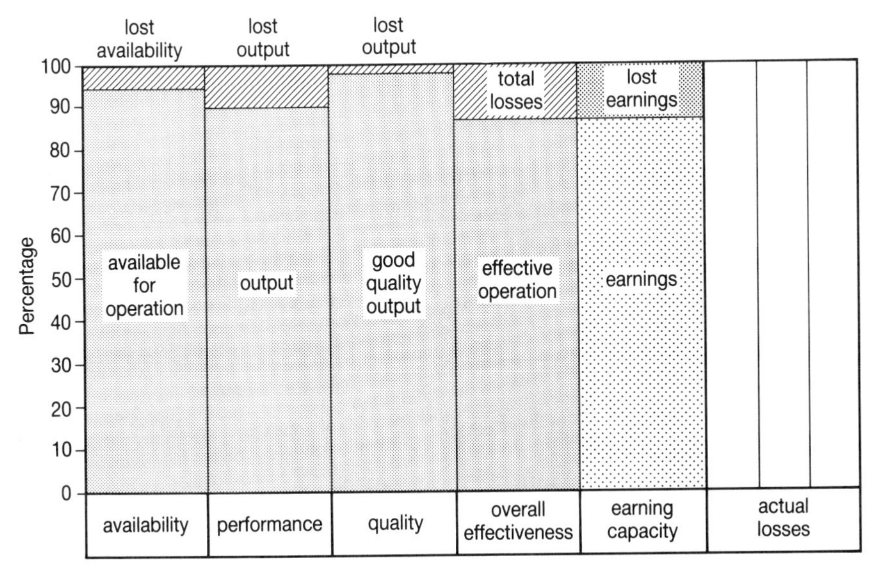

Figure 3.9 Effectogram for example 4

Overall effectiveness is a key, factory floor level measure of performance which can be directly related to the turnover and profit generated by each facility. It provides a means of measuring the main operational losses and, more importantly, monitoring progress in reducing these losses and improving machinery effectiveness. TPM considers overall effectiveness as one of the most significant measures of performance and improvement and the process of monitoring and plotting this alone can provide benefits. Overall effectiveness should not be used just as a means of comparing one operating company with another or of comparing one area of a business with another. It is natural that this comparison will be made and an element of competitiveness will creep in, but the real use of overall effectiveness is as a measure of progress and an indicator to direct improvement activity.

Chronic and infrequent losses

Infrequent or 'sporadic' losses are the obvious losses which occur due to breakdowns, but there are also 'chronic' losses which are the underlying causes of poor effectiveness. Because these losses are not so obvious they are often accepted as the normal operating situation and have the following characteristics:

- these are underlying losses;
- there is a negligible loss per incident;
- it is a major cumulative loss;
- it is a frequent occurrence;
- these issues are often easily restored by operators/craftsmen;
- they rarely come to the attention of supervisors;
- these losses are difficult to quantify;

  they are detected by comparing current conditions to the design or optimum.

Chronic losses contribute to all of the six big losses described in the previous section and are often much more difficult to detect and measure. It is important, however, that chronic losses are considered as they can contribute up to 50 per cent of the losses encountered. The elimination of chronic losses is likely to be the subject of TPM continuous improvement activities.

■ 3.5 Establish a clean and tidy workplace

This component of TPM could be described as 'good housekeeping' and is known to the Japanese as the five S's. The five S's can be roughly translated as:

Cleanliness
Arrangement

Neatness
Discipline
Order

CAN DO disciplines underpin a TPM programme and often provide the most tangible and visual indication that TPM is being implemented in an operating company. Let us explore each of the elements of CAN DO in more detail:

Cleanliness

There is a great deal of emphasis placed on cleanliness in all companies that have implemented TPM. Every area of such companies is normally spotless, even the main production areas, and they are proof that factories do not have to be dirty. Keeping facilities, the workplace and the general environment clean has the following advantages:

- Dirt and debris are not able to cause wear, corrosion and premature failure of machinery and equipment parts.
- The workplace is a much more pleasant place to work and morale is improved.
- Any oil leaks, spillage or abnormal conditions are very easily detected.
- There is a psychological effect that improves people's reactions and performance.
- The workplace is safer as there is a reduced number of hazards, and instructions and warnings are seen more easily.

Cleanliness is the first indicator that the process of change is well underway.

Arrangement

The effectiveness of the process accomplished by machinery and equipment will be improved if all of the facilities, including tooling, fixtures, hand tools, materials, etc. are arranged for ease of access and use. There is an emphasis on improving the layout and position of every aspect of the workplace so that difficult tasks can be eliminated if at all possible. Better arrangement of facilities brings substantial benefits to the people who operate and set up machinery.

Neatness

Having improved the arrangement of tooling, fixtures, hand tools and materials, it is now necessary to store them neatly. Neatness is all about having a place for everything, and a defined location is required for every facility that is used to carry out the production process. Racks, shadow boards, cabinets with special drawers, cupboards, etc. are used to store smaller items and larger facilities such as lifting equipment, vacuum cleaners, etc. have a clearly marked 'parking spot' on the factory floor. By having a clearly defined place it is much easier to encourage people to return the facilities after use and also much easier to check whether the facilities have been returned.

Discipline

Following on from the arrangement of facilities for ease of operation and access and the neat storage of these, discipline is required to ensure the following:

- Facilities are returned to their proper location after use.
- Facilities are kept clean and tidy.
- If facilities are damaged then they are repaired or replaced.

This is all about 'everything in its place' and will stem from the work done in arranging everything neatly and providing 'a place for everything'. More than anything else, discipline stems from the culture within the company and how much everyone cares about the workplace and the general environment.

Order

The maintenance of a clean, neat and tidy workplace will be achieved only if the discipline is in place, and if simple procedures and mechanisms are implemented to support this. In practical terms this means that detailed working procedures are available which remind people of the need to clean, check, tidy up and report any problems. In this way order and control are established in the workplace and the CAN DO activities will be carried out. Also, the use of very obvious visual methods of control such as tags, charts, notice boards and checklists contibutes to the process of change.

More checks, control mechanisms and support are required in the early stages of introducing CAN DO until it becomes second nature and part of the company culture.

■ 3.6 Identify and eliminate inherent faults

As a result of the autonomous maintenance activities the TPM team will discover inherent faults in either the design and construction of machinery and/ or in the methods of operation which support the process. Quite often, they will not be able to rectify these faults as part of the everyday TPM activities as these faults require a great deal of effort, resources and expenditure. Typically, such inherent faults will be limiting the overall effectiveness of the machinery and, although the TPM team will try to reduce their effect wherever possible, a major problem may still exist. This is the point at which the team will propose that a project be set up which will focus on the inherent fault. A small project team, comprising personnel with the appropriate skills to analyse and fix the fault, needs to be set up and the project justified in financial terms using the estimated improvement in overall effectiveness. Figure 3.10 shows the types of focused project which may be required and how they will influence overall effectiveness.

The types of focused improvement project include:

Availability improvements through:

- Changeover and set up reduction projects which analyse all aspects of the activities needed to change from product to product, with the aim of reducing the time and effort required.
- Reliability improvement projects which seek to improve or replace machinery elements which are not robust and reliable.
- Maintainability improvement projects which aim to make the maintenance of replacement machinery elements easier and faster by re-designing and/ or re-positioning them.

Performance improvement through:

- Chronic loss analysis and improvement projects that investigate the losses which reduce the performance of the machinery, analyse and prioritise them and engineer solutions to reduce or eliminate them.
- Process improvement projects which are aimed at improving the robustness and speed of operation of the process, including the investigation of new technology which could be applied.
- Operational improvement projects which concentrate on the methods of operation, procedures, tools and systems with the aim of improving the operator input to the process.

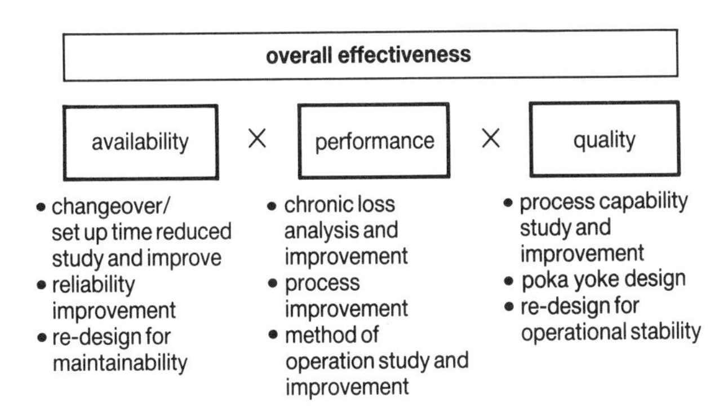

Figure 3.10 Types of focused improvement projects

Quality improvement through:

- Process capability studies which analyse the process capability and machinery capability, investigate variations and seek to eliminate them with the aim of developing a more capable and consistent process.
- Poka yoke (mistake proofing) projects which are aimed at eliminating the possibility of producing defects and making mistakes. Work stations are re-designed to ensure that either mistakes cannot be made or if they are, they are instantly detected.
- Operational stability projects which investigate the reasons why adjustments have to be made to the machinery during operation and seek to make the machinery/process more stable and robust and eliminate the need for these adjustments.

The method of setting up and implementing such projects will be discussed in more detail in Chapter 5.

■ 3.7 Provide maintenance systems to support facilities

The activities of the TPM team in the establishment and restoration of machinery condition and the development of local or 'autonomous' maintenance will have the effect of reducing the number of machinery breakdowns by something in the region of 70 per cent. As illustrated in Figure 3.3, the role of the maintenance department will gradually change from reactive to pro-active and, thus, a maintenance system will be needed to organise and co-ordinate the use of resources and introduce more professional maintenance tools and techniques. It will be useful to understand the various maintenance tools and techniques prior to discussing maintenance systems.

The pattern of maintenance

As discussed in Chapter 2, a detailed analysis of the pattern of maintenance activities employed within many operating businesses would probably indicate that the case is as follows:

- Over 60 per cent of activities were involved in reactive or breakdown maintenance work where production had stopped.
- Around 20 per cent of activities were involved in reactive maintenance work at the end of the shift.
- Only around 15 per cent of maintenance activities were in pursuit of planned maintenance to prevent breakdowns.
- 5 per cent or less of activities were in pursuit of predictive maintenance.

Maintenance staff are, therefore, used to a very reactive method of working, of being 'the hero with the spanner' who enables production to resume again. Figure 3.11 illustrates the pattern of maintenance often encountered within operating companies.

What about the future pattern of maintenance? The pressures which operating companies find themselves under will undoubtedly demand a change in the pattern of maintenance to probably less than 10 per cent reactive maintenance which disrupts production, and the majority of activities being in support of planned preventive and predictive maintenance.

Maintenance techniques

We shall now discuss the following maintenance techniques:

1. Reactive (breakdown) maintenance
2. Planned, preventive maintenance
3. Reliability-centred maintenance
4. Predictive (condition based) maintenance

1. Reactive (breakdown) maintenance

As previously mentioned, reactive maintenance is the traditional and most widely used approach within manufacturing industry, despite the following characteristics:

- No warning of failure with possible safety risks.
- Unscheduled down-time of machinery.
- Production loss or delay.
- The need for standby machinery where necessary.
- The need for maintenance resources to be instantly available.
- Possibility of secondary damage.
- Problems in obtaining spare parts at short notice or having to hold a substantial stock.

Reactive maintenance can be cost-effective in some situations, particularly

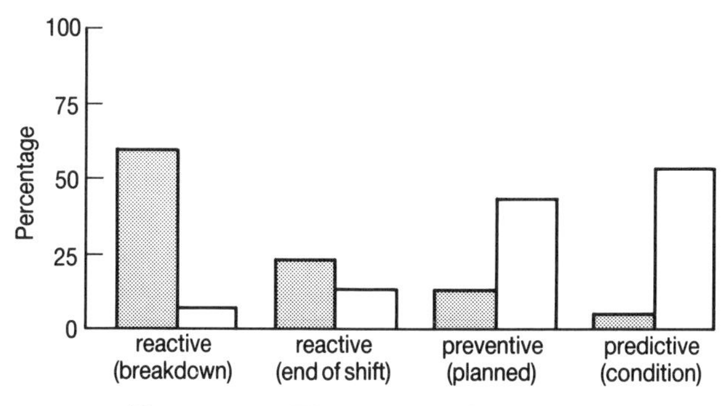

Figure 3.11 The pattern of maintenance

where machinery is not critical to production, simple and/or inherently reliable.

2. Planned preventive maintenance

Planned preventive maintenance provides such advantages as follows:

- Greater control and confidence in the availability of machinery and equipment.
- Scheduling of maintenance tasks with production requirements.
- Scheduling of maintenance resources to give the most efficient use of labour.
- Avoidance of the premature deterioration of machinery.

The cost effectiveness of planned, preventive maintenance does, however, depend upon the following:

- The knowledge of machinery and its failure/deterioration characteristics.
- The quality and effectiveness of inspections, where these are carried out.
- The correct time intervals between replacements and the quality of the work carried out.
- The discipline of keeping records and continually updating the system.

Planned, preventive maintenance does have the following disadvantages:

- It can increase maintenance costs significantly through set up and over-maintaining machinery.
- Sometimes it can cause failures as a result of the planned maintenance activity.
- It can only really be effective where the deterioration is age-related.

3. Reliability-centred maintenance

Reliability-centred maintenance (RCM) is an analytical approach to help prioritise maintenance tasks and machinery. It can be used to concentrate effort where it is really needed and so make the most effective use of resources. RCM uses information from experienced operators, craftsmen and maintenance staff and operates through the use of analysis techniques such as the following:

- Failure modes effect analysis (FMEA).
- Cause and effects analysis.
- Risk analysis.

It thus identifies specific maintenance tasks and the frequency with which they need to be carried out.

4. Predictive (condition-based) maintenance

A desirable situation would be to predict a failure before it occurs, with enough time to be able to make arrangements to replace the faulty part prior to failure and without affecting production. Predictive maintenance aims to do this, and a number of condition-monitoring techniques are used such as:

- Thermography which is used on cables and switch gear to show up hot spots. This produces an infra-red image of the subject which can be recorded by photograph or computer.
- Acoustic emission, a well-established technique for the detection of cracks and weld faults in pressure vessels. It can also be used for monitoring low-speed bearings and uses transducers attached to the subject.
- Oil analysis which examines lubricating oil for signs of wear, contamination, etc.
- Vibration analysis which uses an accelerometer to measure vibration which is processed and presented in the form of a graph. The graph is integrated and transformed to a spectrum over a frequency range.
- Monitoring other process parameters such as, pressure, flow, temperature, load and wear.

Predictive maintenance is not always appropriate but where it is, it permits the shutdown of machinery before any damage occurs, allows maintenance work to be planned and organised more efficiently and production to be reorganised accordingly.

Development of maintenance techniques

There are quite obvious advantages and disadvantages with all of the approaches described, and it would be wrong to recommend any as the best for all businesses. For particular applications within particular businesses, it may be that one approach is most cost-effective and then that should be employed. The achievement of 'best practice' can be achieved by the installation of an appropriate maintenance system which uses the most cost-effective maintenance approach or approaches for the business, combined with localised, operator-led 'autonomous maintenance' activities, all implemented as part of an overall TPM programme. Thereafter, further improvements will be achieved by designing out the maintenance requirements of existing and new machinery.

Through the TPM teams, simple yet detailed procedures will be developed for keeping the machinery in good condition. Some of the tasks which are detailed in the procedures will be allocated to the maintenance department as more specialist skills, experience and/or tools are needed to accomplish them. These tasks will form the basis of a preventive maintenance schedule for the machinery and, when consolidated for a given area of the factory, will form a schedule for that particular area. The list of activities, requirements and time intervals will be meaningful and easily understood because they have been generated by the people who operate, set up and maintain the machinery and have not been imposed by an external party.

The schedule can be further developed and enhanced as TPM progresses and more regular servicing and opportunities for machinery condition monitoring will be identified.

■ 3.8 Purchase and install facilities that provide the best return

Operating businesses are often faced with the need to improve existing manufacturing facilities or install new facilities which includes machinery of various types. It is very important that investment decisions are based upon the needs of the business if maximum benefit is to be extracted from the investment. Traditionally, the selection of machinery for a business has been based upon the degree of technical excellence of the machinery in performing a specific function along with the cost, with little regard to the overall business needs or compatibility with the associated manufacturing systems, environment and operator needs.

Manufacturing businesses need to invest in machinery and associated manufacturing facilities in order to compete in the world marketplace and, therefore, any investments made now or in the future are even more critical than has been the case previously.

The selection and purchase of machinery must be approached in a professional and structured way, from initial consideration right through to the achievement of the desired production levels. The machinery must match the overall business strategy and provide real benefits to the business. Traditionally, the approach to the selection of machinery has been to address each item in isolation and seek technical excellence in specific areas of function and performanc‘ w‘‘‘‘‘‘‘‘ay be argued, little regard to the wider business needs or compatibility with other associated elements of the manufacturing system. This approach does not make effective use of the capital invested by the business and can often be shown to adversely effect the performance of the business and place constraints upon future improvement plans.

In order to meet the present and projected future needs of the business, it is important that all machinery applications are considered as an integral part of the overall manufacturing system. The machinery installed must be compatible with other associated systems, conform to the TPM philosophy of the company, must provide appropriate interfaces with associated systems and also consist of elements which are compatible with each other. Thus, there is a fundamental need for the application of a structured, systems approach to the planning and procurement of machinery tools as part of an overall objective to achieve world-class competitive performance in manufacturing.

In addition to hardware and software compatibility other factors should be considered such as follows:

- Manufacturing control systems.
- Quality requirements.
- People skills.
- Level and type of resources available.
- Autonomous maintenance activities.
- Maintenance systems.
- Appropriate level of technology.

These are a necessary input to the selection of machinery which will meet the needs of the business. Figure 3.12 illustrates these factors.

Machinery life-cycle costs

It has been estimated that around 70 per cent of life-cycle costs are fixed by the end of the concept stage of machinery design. Figure 3.13 illustrates the activities which affect life-cycle costs.

The life-cycle cost of machinery is the total cost of ownership and can be divided into the following:

- Acquisition cost. The capital cost plus cost of delivery and installation.

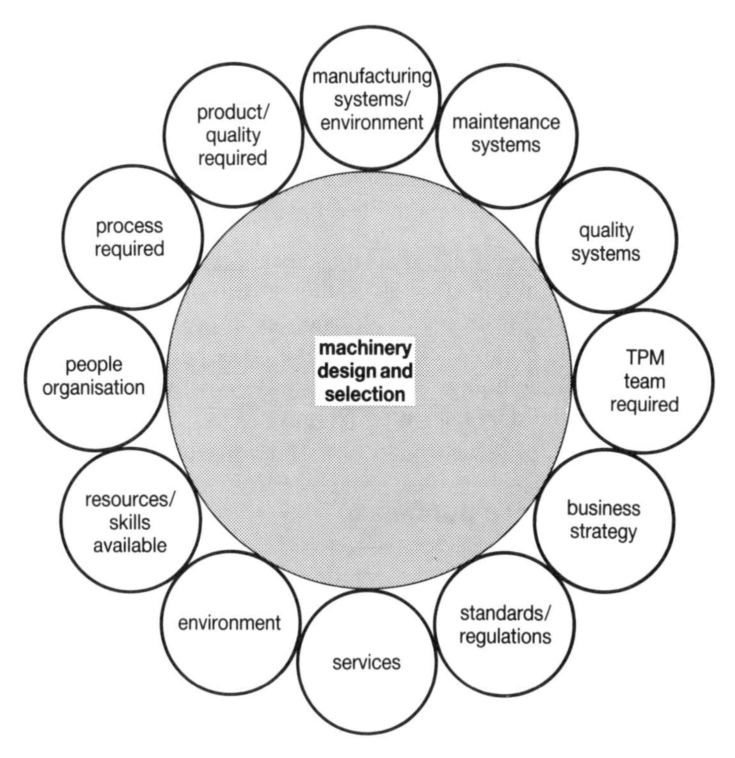

Figure 3.12 Factors affecting machinery design and selection

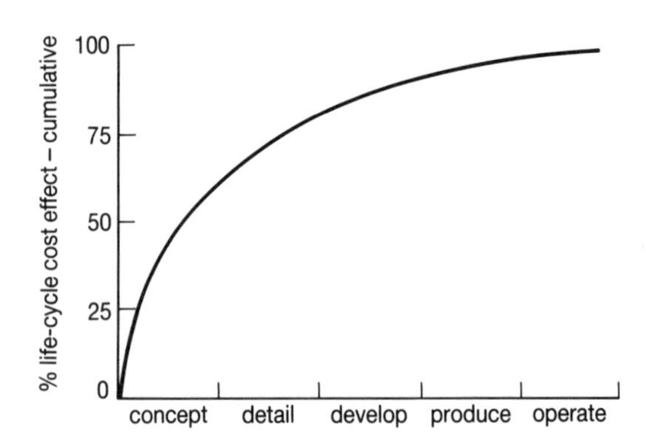

Figure 3.13 Activities affecting life-cycle costs

- Ownership cost. The cost of modifications, preventive and corrective maintenance.
- Operating cost. The cost of materials, consumables and energy.
- Administration cost. The cost of data acquisition, recording and documentation.

It is important to consider life-cycle costs when planning for a new machinery investment as a lower capital cost will not necessarily mean lower overall life-cycle cost.

Life-cycle costs will be directly influenced by the following factors:

- Reliability which determines frequency of repair, fixes spares requirement and loss of revenue due to lack of availability.
- Maintainability which affects the level of resources and skills required.

The manufacturer's pre-delivery costs i.e. design, procurement and manufacture, increase with availability. This is due to the additional design effort, selection of higher specification parts, higher levels of testing, etc. required to build in reliability and maintainability. On the other hand, the manufacturer's after-delivery costs including warranty, re-design, modification, loss of reputation, contractual damages, etc. decrease as availability is improved.

Using LCC to influence purchases

From the user's point of view the relationship between the capital cost of machinery and the costs of failure for that machine can be shown to have an optimum level of availability which incurs minimum cost. This attempts to illustrate that life-cycle costs can be minimised as a result of efforts to obtain reliability and maintainability enhancements, even if higher capital costs result. Life-cycle costs are unfortunately not used very much in industry to justify the purchase of machinery, the emphasis being placed upon the capital cost only.

There is a tendency, however, to select machinery which is perceived to be more reliable by virtue of the manufacturer's reputation or user experience. This in effect takes into account implied life-cycle costs, and information should be available regarding the capital and operating costs of 'standard' machinery where historical records have been compiled. By adding the capital costs of the machinery, the operating costs of day-to-day running and the ownership/support costs of maintaining, replacing parts, inspection, and modification, etc., an overall life-cycle cost can be established. This enables a comparison to be made between different manufacturers and items of machinery.

For new, special purpose or machinery for which information is not available, it will not be possible to establish detailed ownership (support costs), but it should, however, be possible to gather information pertaining to capital cost and estimated operating costs from the manufacturer who can often be persuaded to estimate replacement parts usage and cost, inspection and preventive maintenance frequency and periods. Thus an estimate of the life-cycle costs of hitherto unknown machinery can be made. This can be used to justify the purchase of machinery which may be more expensive initially, but overall will offer lower costs to the business.

The whole cycle of identifying the need, planning, selection and implementation of a machinery investment can be regarded as a project in its own right, although it may be a work package within a much larger project aimed at the reorganisation and improvement of existing facilities or resulting from a new product introduction. More details of the stages of machinery purchase including specifications are provided in Chapter 5.

4 The Benefits of TPM

We have established that TPM enables maintenance systems to be developed and operated effectively and integrates the activities of production and maintenance personnel. It brings a different approach to both operations and maintenance and encourages a change in attitude which benefits everyone within the company. Working within a reactive maintenance, 'I operate, You fix' environment can be very stressful to production and maintenance people alike. It breeds despondency, apathy and a feeling of not being in control of the work/workplace and it leads to a great deal of waste.

TPM has been tried and tested over many years, in many industry sectors and in many parts of the world, and an overwhelming body of evidence proves that it does work. It makes businesses more competitive, changes the working environment for the better, encourages and enhances people and, overall, transforms the way in which a company operates.

■ 4.1 Enhancing production personnel

Many traditional operating methods and maintenance approaches have tended to assume that factory floor personnel have very limited intellectual capacity or integrity and, as a consequence, have to be 'managed' very closely. There has been a great deal of effort to de-skill people's jobs, reducing them to boring, repetitive activities. This, coupled with the requirement for efficient operation producing the maximum output in the shortest possible time, has created an atmosphere of mistrust and sometimes confrontation and has led to a de-motivated workforce.

As discussed in Chapter 2, factory floor personnel are often blamed for poor manufacturing performance even though they are poorly trained, de-motivated and provided with machinery which is well below standard. TPM recognises that factory floor personnel are a key business asset and that they should not be wasted, thus the whole TPM approach is based upon the motivation, participation and enhancement of shop floor personnel, allowing them to have some control over their own working environment.

Factory floor people are encouraged to look after 'their' machinery and 'their' equipment/tools and to actively participate in the improvement of production methods and processes. They are trained and developed to realise their full potential and rewarded for their contribution towards making the business more competitive and profitable. TPM brings considerable benefits to factory floor personnel, both in terms of an improved working environment and enhancement of their skills and personal attributes. Figure 4.1 illustrates how TPM benefits machinery operators and craftsmen. The main benefits for operators are as follows:

- A clean, tidy and safer workplace. One of the most tangible benefits of TPM is the improvement in the workplace and general working conditions.
- Problems and faults fixed. Through the TPM team, the problems and faults which have hindered people's jobs and fuelled frustration are rectified.
- A say in what goes on in their area. More control over their own working environment, opportunities for suggesting improvements and getting the changes that they really want.
- Opportunities to increase skills and knowledge. By working more closely with craftsmen, technicians and engineers a great deal is learned about machinery, production and engineering principles. TPM programmes incorporate increased training and development opportunities for factory floor personnel.

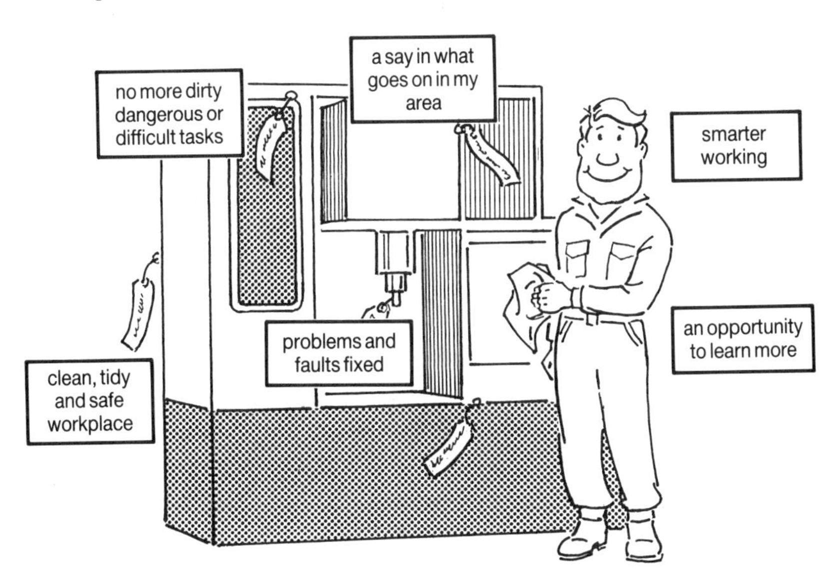

Figure 4.1 The benefits for operators and craftsmen

- Less panic and more control. Because faults and operating problems are fixed properly, the working environment becomes more controlled and pleasant for all concerned.
- Smarter methods of working on more effective machinery. All dirty, dangerous and/or difficult tasks are investigated and eliminated if at all possible.

It is my experience that these benefits become evident very quickly and that once factory floor personnel have embarked upon a TPM programme they do not want to return to the previous method of operation.

4.2 The benefits for maintenance personnel

What are the benefits for maintenance people who have to fix equipment when it breaks down and try to keep production going? The benefits for them of TPM are as follows:

- Machinery, equipment, tooling, etc. which is clean and kept in good condition and, as such, is better and safer to work on.
- Less breakdown maintenance (reactive work). Through TPM activities the number and frequency of breakdowns is substantially reduced and thus the amount of time spent in reacting to panics is much less.
- Less time spent on simple, unskilled jobs. Skilled maintenance technicians and engineers are relieved of the many simple, unskilled tasks that they have traditionally carried out.
- More time to spend on preventing breakdowns. Because there are less breakdowns and the need to carry out simple, unskilled tasks has been removed, there is more opportunity to do what maintenance personnel are best equipped to do, i.e. preventing and predicting breakdowns and machinery deterioration.
- More time to spend on getting rid of the causes of machinery and equipment problems. Similarly, there is more opportunity to properly analyse the cause of breakdowns and reduced performance and come up with engineering solutions.
- Opportunities to increase skills and knowledge. As the role of maintenance personnel changes, they will learn about other engineering and maintenance tools and techniques and how to apply them.

Maintenance personnel also gain from a closer working relationship with production and become involved with training and advising production personnel.

4.3 The quality dividend

Maintenance is traditionally associated with machinery reliability as measured by availability or up-time and the connection between machinery condition and product quality has not really been recognised. TPM, with its emphasis on improving all aspects of machinery operation and eliminating all losses, definitely links machinery condition with the quality of products. In almost all instances where TPM has been implemented within manufacturing businesses it has led to improvements in product quality, which can be measured by the number of defects produced, amount of re-work and/or yield figures. Some of these improvements have been quite remarkable and prove that process capability can not only be achieved but also maintained. TPM measures of performance and improvement mechanisms identify quality problems and ensure that they are addressed and the pursuit of total quality is inherent in the TPM approach.

As discussed in Chapter 2, TPM is the manifestation of total quality on the factory floor, at the 'sharp end' where the wealth is created by an operating business. Importantly, anyone involved in implementing TPM will be able to experience physical changes in the working environment which can happen very quickly along with a profound change of attitude from everyone in the company.

■ 4.4 Business improvements

TPM is an approach which provides benefits to the whole business in the form of:

- Improved effectiveness of machinery and equipment which directly affects key business ratios and competitiveness.
- Improved quality of products, less scrap and re-work which not only reduces manufacturing costs but increases customer satisfaction.
- Enhanced factory floor personnel, improved motivation and morale, arising from a much improved working environment, greater participation and training.
- A more controlled and well-organised manufacturing operation with less pressure and 'fire fighting' and more time for continuous improvement and development.
- A much better working environment for everyone.

Some examples of the overall business improvements achieved as a result of TPM implementation are given below:

Company A glass manufacturer
- Breakdowns in factory reduced from 150 to 10 a month.
- Throughput increased by 40 per cent.
- Defects reduced by 30 per cent.
- Overall equipment effectiveness of 86 per cent achieved.

Company B automotive supplier
- Breakdowns in factory reduced from 800 to 5 a month.
- Defects reduced by 60 per cent.
- Overall equipment effectiveness of 80 per cent achieved.

Company C automotive manufacturer
- Breakdowns in factory reduced from 1800 to 170 a month.
- Defects reduced by 90 per cent.
- Throughput increased by 30 per cent.
- Energy costs reduced by 25 per cent.

Company D food manufacturer
- Breakdowns in process area reduced from 300 to 100 a month.
- Throughput increased by 25 per cent.
- Overall effectiveness of packaging machinery increased from 62 per cent to 80 per cent.

These benefits were not achieved overnight; in some cases TPM had been running for several years but from the early stages of applying TPM, improvements were achieved and continued to provide substantial benefits to the businesses. Any company which applies TPM in a thorough and committed manner can expect to achieve the benefits described in this chapter, especially when allied to the effective application of other maintenance tools and techniques.

4.5 Transforming the typical scenario

In Chapter 2 we described the typical scenario that is found in many operating companies and listed the characteristics of how they operate. So how does TPM transform this scene? Well, first, the company must acknowledge the truth of its present situation and wish to change. TPM then addresses each aspect of the operation of the company as follows:

- The combination of production-led, autonomous maintenance activities, focused improvement projects and effective maintenance systems (as discussed in Chapter 3, sections 3.2, 3.3, 3.6 and 3.7) will greatly reduce the number of machinery breakdowns.
- Through monitoring, plotting and analysing overall effectiveness information (as discussed in Chapter 3, section 3.4) the significance of product changeover and adjustment times will be established quickly. The overall effectiveness information will help to prioritise and justify the need for and cost of changeover time reduction projects and will provide the impetus necessary to carry out such projects. Under the auspices of TPM changeover time reduction projects will be carried out and, if carried out in a structured way (as detailed in Chapter 5, section 5.5), reductions of at least 50 per cent will be achieved.

- As a result of autonomous maintenance activities and improved maintenance systems, the general condition of machinery will be improved. This will lead to an improvement in the quality of products and the confidence of personnel in the consistency of the process. Also, where the overall effectiveness information can be used to identify problem areas and justify focused improvement projects such as process capability studies, these will improve the consistent quality of products.
- TPM activities will gradually rebuild employee morale and relationships. By improving not only machinery but also the general working environment, factory level personnel will benefit, and morale and enthusiasm will improve. Communications will greatly improve as frank and honest 'two-way' communications mechanisms are put in place through the TPM teams.
- Many TPM activities subtly bring together people from many levels of the company and from different departments. Once problems are being approached in a professional and structured way then the true situation and cause of problems can be clearly identified. TPM insists that a 'no blame' culture is cultivated and, by working together, the traditional barriers between departments and levels within the company are gradually dismantled.
- Panics gradually become a thing of the past and personnel are taught new skills which are more related to planning, analysing and improving rather than reacting. When people start to operate within a more controlled and organised environment they become determined not to allow the situation to revert to 'panic mode' again.
- The working environment becomes much cleaner, tidier and neater and standards of cleanliness and tidiness gradually become higher and higher. One very noticeable characteristic of any TPM factory is that it is very clean and tidy and this in itself will improve people's morale and pride in the workplace.

Overall, the normal working day becomes less stressful and reactive and much more controlled and as one production manager, whose company has made the transition, described the new scenario, 'I don't have to run the factory floor any more, the teams do and, therefore, I can dedicate a great deal of my time to supporting their improvement activities and planning for the future.'

■ 4.6 Justifying a TPM programme

To whole-heartedly implement TPM across the company takes time, effort, financial and human resources but very substantial benefits will result. In this chapter we have already described many of these benefits, the most significant of which are the improvement in morale, enthusiasm and the release of the previously untapped potential of everyone within the company. This particularly applies to factory floor personnel.

I believe that if morale within the company is good, then the company will be successful and will overcome any difficulties that may arise as a result of changes in the external influences described in Chapter 2 (illustrated in Figure 2.2). In the 'real world', however, the majority of companies operate fairly stringent financial controls, which require that any expenditure has to be justified on the basis of a return on investment. Money is usually in short supply in operating companies and there will be a number of 'improvement initiatives' which are competing for limited funds. In my experience, many senior managers prefer to spend company funds on capital investment projects rather than these 'improvement initiatives' as they are more tangible and more easily justified. It is necessary, therefore, to be able to justify a TPM programme in financial terms and to be able to show a return on investment.

The cost of implementing TPM

If the financial justification is to be compiled, then first the cost of implementing a TPM programme has to be estimated. It is very difficult to estimate the overall cost of implementing TPM without first of all running one or more TPM pilots. More details of setting up and implementing TPM pilot areas are provided in Chapter 5. As a result of implementing a pilot area or areas, the true cost of implementation across the whole company can be estimated. The major cost elements will probably include the following:

- The direct cost of the TPM team members – man hours applied to TPM activities.
- The direct cost of other company personnel (such as maintenance people) – man hours applied to TPM activities.
- The cost of any external resources required such as sub-contract labour and service engineers.
- The cost of spare parts, materials and other items which have to be purchased.
- Training and consultancy support costs.
- Lost production revenue resulting from the need to stop the process whilst TPM activities take place.

These costs can be recorded for the pilot implementations and used as a basis for estimating the cost of spreading TPM across the company.

The financial benefits of implementing TPM

In Chapter 3 the concept of the six big losses and overall effectiveness was introduced and the 'effectograms' in Figures 3.5 to 3.9 illustrated the effects of the losses on the earning capacity of the particular machinery or process. Overall effectiveness is a direct measure of the earning capacity of facilities within the company and can be used to measure the financial benefits arising from the application of TPM.

The processes which are employed within an operating company are designed to add value to the materials or parts which they are processing. For example, if a press is used to press out motor car body panels from a sheet of steel, the value of the processed panel will be greater than that of the blank sheet. The difference between the two is termed 'added value' and is calculated as follows:

cost of blank sheet = £5.00
value of pressed panel = £7.00
added value = £7.00 − £5.00 = £2.00

Thus, the process has earned £2.00 for the company. If the expected throughput is 50 panels every hour then the added value per hour can be calculated as:

added value/hour = £2.00 × 50 = £100.00/hour

The expected throughput will probably be based upon the theoretical cycle time for the process and will not take into account the six big losses which were explained in Chapter 3. If these losses are recorded and a figure for the overall effectiveness of the press calculated as, say, 70 per cent then the actual added value per hour can be calculated as:

$$\text{actual added value/hour} = £100.00 \times \frac{70}{100} = £70.00\text{/hour}$$

and therefore the six big losses represent a loss of added value equal to £30.00/hour.

An annual loss figure can be determined by estimating the average loading of the press during the year. For instance, if the press was used on a single shift basis and operated on average 35 hours per week and for 48 weeks per year then the average annual loading can be calculated as:

average loading = 35hrs × 48 wks = 1680 hours/year

therefore, the annual loss figure is the product of the loading hours and loss per hour:

annual loss = £30 × 1680 hours = £50 400/year

This means that the press could have earned an additional £50 400 for the company if it operated at 100 per cent effectiveness rather than 70 per cent effectiveness. By taking this figure, which represents the lost earning capacity of the machinery, and dividing it by the overall effectiveness loss, a figure for the additional earning capacity gained for each 1 per cent improvement in overall effectiveness can be calculated, thus,

100% − 70% = 30% loss in effectiveness = £50 400 loss in earning capacity, therefore

$$1\% \text{ improvement} = \frac{£50\ 400}{30\%} = £1680/\text{per cent}$$

For this example, every 1 per cent improvement in overall effectiveness will provide an additional earning capacity of £1680 per year.

In my experience, the introduction of autonomous maintenance activities as outlined in Chapter 3, sections 3.2, 3.3, 3.4 and 3.5, which are all production-led activities, will lead to a rapid improvement in overall effectiveness of at least 5 per cent, but more usually of the order of 10 per cent. Therefore, for the example of the press, autonomous maintenance activities would be expected to increase its earning capacity by at least £8400 and most likely the figure would be £16 800.

Most autonomous maintenance activities, as Chapter 5 will explain, do not involve a great deal of expenditure. The major costs involved are people costs and typically for an exercise on machinery such as a press the costs may be as follows:

The direct cost of the TPM team members = 130 man hours @ £10.00/hour = £1300.
The direct cost of other company personnel = 30 man hours @ £12.00/hour = £360.
The cost of any external resources = £250.
The cost of spare parts, materials and other items which have to be purchased = £500.
Share of training and consultancy support costs = £2500.
Lost production revenue = £2800 (40 hours stopped @ £70/hour).
Total costs = £7710

If the TPM activities could have been scheduled to take place during a production 'window' then the lost production revenue figure would be greatly reduced.

The financial justification for undertaking the launch of TPM activities on the press can now be based upon these figures which would show a payback calculated as follows:

$$5\% \text{ improvement} = \frac{£7710}{£8400} = 0.92 \times 12 = \text{approximately 11 months}$$

$$10\% \text{ improvement} = \frac{£7710}{£16\ 800} = 0.46 \times 12 = \text{approximately 5.5 months}$$

This calculation is based purely on the additional earning capacity which will be gained as a result of improved overall effectiveness and does not take account of the savings which will be made as a result of less panic costs including the need for premium time working, subcontracting and the disruption of schedules and a reduction in reactive maintenance costs and 'safety' costs as illustrated in Figure 2.11.

Even based upon financial figures alone, the justification for TPM is very sound indeed and a payback measured in months is normal. Many applications of autonomous maintenance that I have been involved in have resulted in improvements of overall effectiveness in excess of 15 per cent which have been achieved in a matter of 6 to 8 weeks and have provided a payback of less than 3 months.

The examples which were shown in Chapter 3, included a list of different types of losses incurred in a particular environment which were measured and then overall effectiveness figures calculated. It is now appropriate to take each example and calculate the losses in financial terms.

Example 1. High volume production

loading = 80 hours/week for, say, 48 weeks/year = 3840 hours/year
value added to each unit = £0.90 on average
planned throughput = 60 units hour

Therefore, the theoretical earning capacity per year can be calculated as:

earning capacity = 60 units/hr × 3840 hrs/yr × 0.90 = £207 360/year
overall effectiveness = 65%
therefore, losses = 35%
= £207 360 × 0.35 = £72 576/year

and for every increase in overall effectiveness of 1 per cent the additional earning capacity can be calculated as:

$$1\% \text{ improvement} = \frac{\pounds 72\,576}{35\%} = \pounds 2074/\text{year}$$

A glance down the list of faults provided in Chapter 3, section 3.4 would suggest that, for this example, there is great potential for reducing or eliminating many of the faults and thus an improvement of 10 per cent in overall effectiveness should be achieved quite quickly. In this case then the financial payback would be obtained in a matter of a few months. Figure 4.2 shows the 'effectogram' for this example complete with the financial figures.

Example 2. Hand assembly

planned loading time = 8 hours/day × 5 days/week for 8 weeks
= 320 hours in total

If each man hour is sold at £25.00 (assembly processes are often priced on the

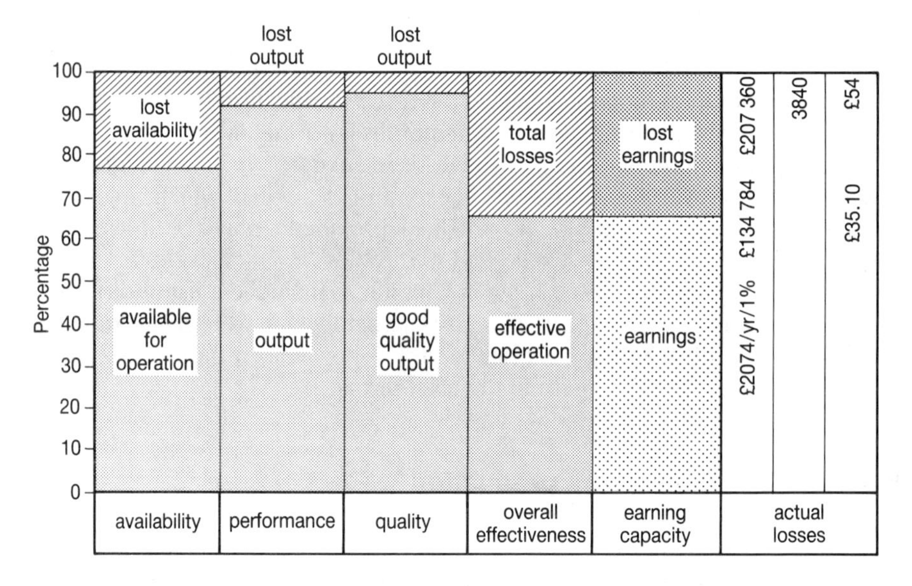

Figure 4.2 Completed effectogram for example 1

man hours required to complete the assembly) then the estimated cost of the assembly can be calculated as follows:

cost = 320 hours × £25.00/hour = £8000
overall effectiveness = 70%
therefore, the actual number of hours required to complete the assembly

$$= \frac{320 \text{ hours}}{0.7} = 457 \text{ hours}$$

A loss of 30 per cent in effectiveness has caused an overrun of 137 man hours which can be calculated as follows:

cost of losses = 137 hours × £25/hour = £3425 for the assembly

If say, 10 such assemblies are carried out every year then the losses = £3425 × 10/year = £34 350/year. The savings which would be achieved for each 1 per cent improvement in overall effectiveness can be calculated thus:

$$1\% \text{ improvement} = \frac{£34\,250}{30\%} = £1142/\text{year}$$

Therefore, in this case a 10 per cent improvement in overall assembly effectiveness would probably provide a payback within 6 months. Figure 4.3 shows the 'effectogram' for this example, complete with the financial figures.

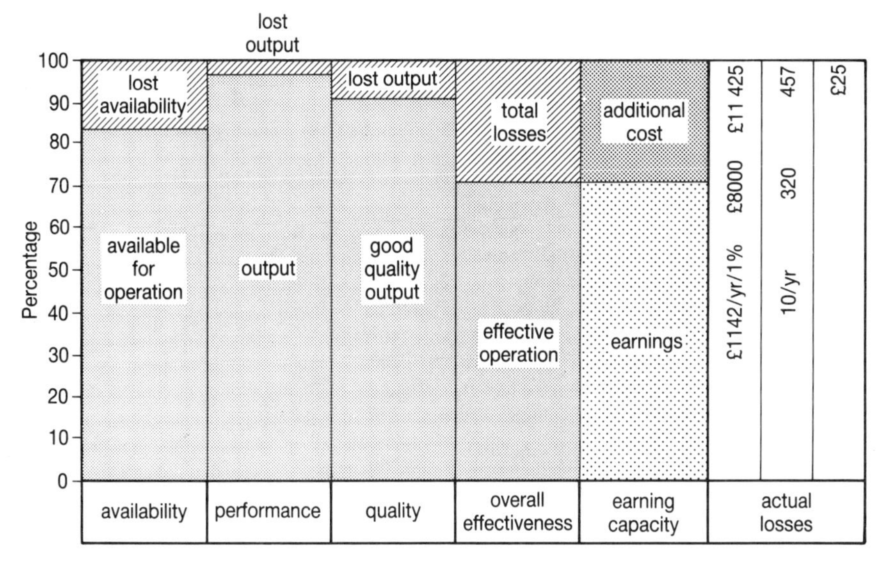

Figure 4.3 Completed effectogram for example 2

Example 3. Semi-automated assembly

loading = 7.5 hours/day × 5 days × 48 weeks = 1800 hours/year
planned throughput = 150 units/hour
value added to each unit = £0.15

Therefore, the theoretical earning capacity for the machinery can be calculated thus:

earning capacity = 1800 hrs × 150 units/hr × £0.15/unit = £40 500/year
overall effectiveness = 42%

therefore, the cost of the six big losses can be calculated thus:

$$\text{losses} = £40\,500 \times \frac{58}{100} = £23\,490/\text{year}$$

$$\text{and 1\% improvement} = \frac{£23\,490}{58\%} = £405$$

In this case, the machinery has a very low overall effectiveness and autonomous maintenance activities would improve this by at least 15 per cent quite quickly. This should provide a payback within 8 months. Figure 4.4 shows the 'effectogram' for this example, complete with the financial figures.

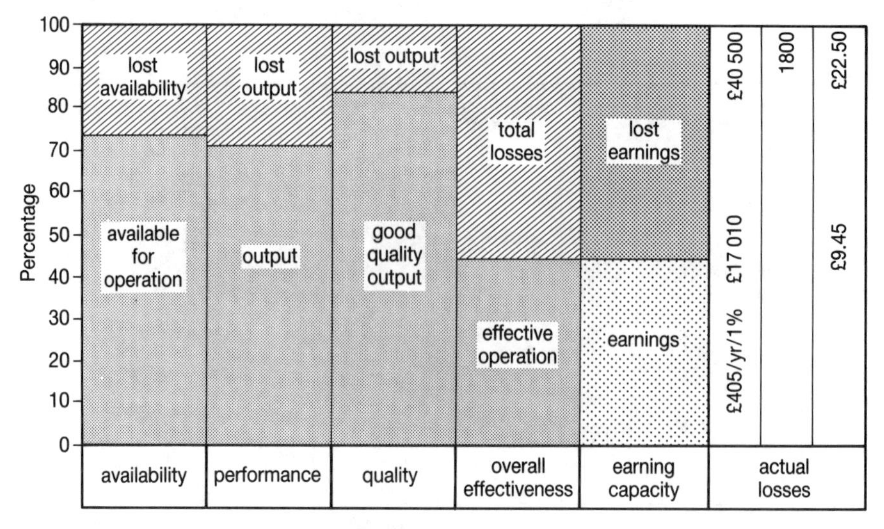

Figure 4.4 Completed effectogram for example 3

Example 4. Processing system

loading = 24 hours/day × 7 days/week × 50 weeks/year = 8400 hours/year
planned throughput capacity = 300 kg/hour
value added to each kg = £0.3/kg

The theoretical earning capacity for the processing system can be calculated thus:

earning capacity = 8400 hrs/yr × 300 kg/hr × £0.3/kg = £756 000/year
overall effectiveness = 85%

therefore, the losses can be calculated thus:

$$\text{losses} = £756\,000 \times \frac{15}{100} = £113\,400/\text{year}$$

$$\text{and } 1\% \text{ improvement in overall effectiveness} = \frac{£113\,400}{15\%} = £7560/\text{year}$$

In this environment the cost of implementing autonomous maintenance may be more substantial due to the type of machinery and the cost of lost production but the potential benefits are greater. An improvement in overall effectiveness of around 5 per cent should be achieved and, therefore, a payback of less than six months would be expected. Figure 4.5 shows the 'effectogram' for this example complete with the financial figures.

The examples have shown how overall effectiveness can be used to financially justify TPM implementation, and this does not just apply to the production led, autonomous maintenance activities but the other components of TPM which will be described in Chapter 5.

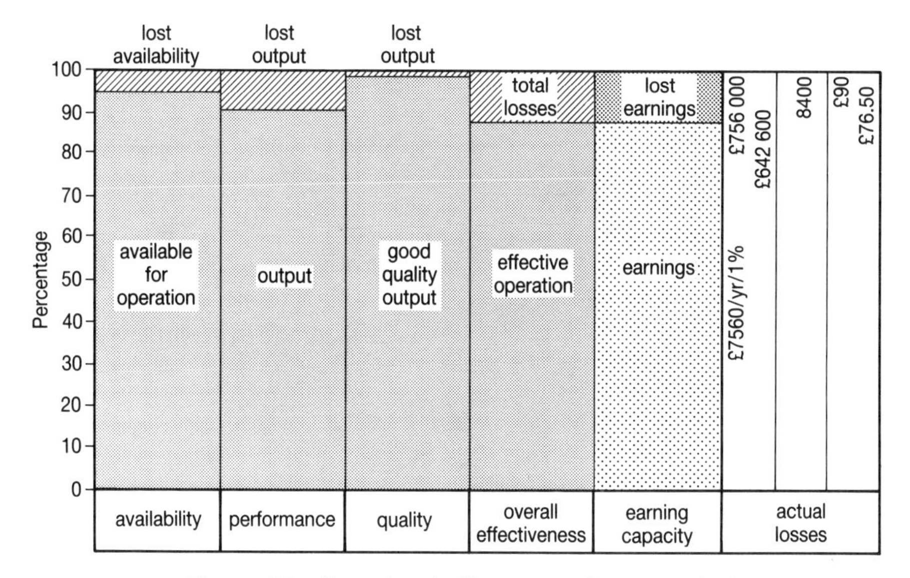

Figure 4.5 Completed effectogram for example 4

5 Implementing TPM

TPM is not costly to implement, it does not require any large capital sums, substantial consultancy fees or licences, but it is not easy and should not be perceived as a short-term measure. It will not work unless the implementation programme is given the wholehearted backing of the management team in terms of actions, not just words. The change of attitude of everyone in the business and the substantial benefits which accompany this may take many years to achieve, but many benefits will be realised from day one of implementation.

TPM is not a difficult concept to understand and its practices and techniques are all quite straightforward and logical. It is, however, the change in attitudes and values accompanying the change in working practices that can be more difficult to grasp and implement. In order to achieve a successful implementation, it is essential that the philosophical changes as discussed in Chapter 3, section 3.1 are achieved at all levels of the company. This means the gradual erosion of many deep-seated, traditional attitudes that have evolved over many years. TPM brings about the change by means of its philosophy, practices and techniques.

The important difference between TPM and some other 'business initiatives' is that it is a very practical concept that brings about immediate and very visible changes to the factory floor environment. This chapter deals with the important subject of implementing TPM from its initial introduction to the company, setting up and running pilots, establishing a long-term programme and initiating the training and development of factory floor personnel.

■ 5.1 Introducing TPM

The key to the introduction of TPM in any business lies in the effectiveness of communications. The principles, techniques and implications of applying TPM within the business must be explained to personnel at all levels via a structured communications programme. It is often advantageous to involve some external, 'independent' party to provide awareness training, presentations, advice and to support the early implementation.

It is usually necessary to choose one or more pilot areas in order to try out TPM, to demonstrate its benefits and establish the practical requirements of implementing TPM on a particular site. When selecting a pilot area it is necessary to consider the following:

- The size and location of the area.
- The amount of machinery and people in the cell/area.
- The type of machinery in the area.
- The enthusiasm of local supervision and factory floor personnel.
- The importance of the area to the business.

The pattern of communications up to the launch of the pilot area, as illustrated in Figure 5.1, has proved to be very effective. Communications may begin with an initial presentation to the executives of the business to 'sell' the concept of TPM. This may be undertaken by a manager from within the company who has gained some awareness of the principles and practices of TPM and has become an enthusiastic sponsor of its application within the company, or by an external party with knowledge and experience of TPM. Having given many executive awareness presentations to senior personnel within many wide and varied businesses, over many years, I have found that reactions vary and that the following are typical:

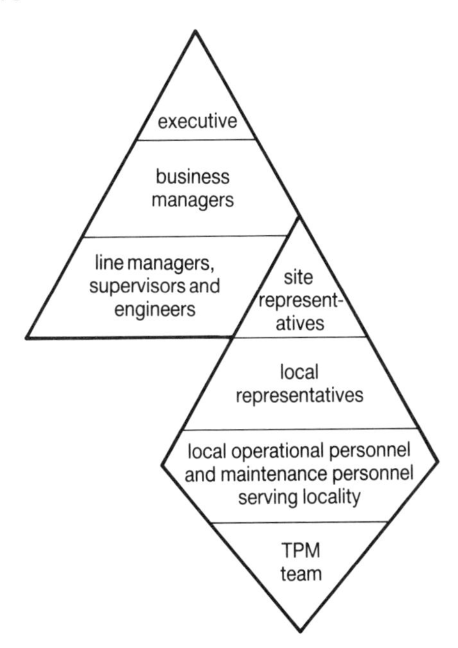

Figure 5.1 The pattern of communications

We are already undertaking a total quality programme and doing all of these things, so what is so different about TPM?
This comment often comes from someone who has been involved in or has sponsored the 'total quality initiative' and is a very defensive reaction. The validity of this statement can soon be checked by visiting the factory floor and talking to factory floor personnel. In the majority of cases the initiative has not been very successful and has not addressed the facilities on the factory floor and has done little to improve morale.

This all sounds very good, but we can't afford the time or resources to carry out a programme of this kind.
Usually this point will be raised by the operations or production manager who is so used to the 'typical scenario' (as discussed in Chapter 2) that he/she is conditioned to operating in 'panic mode'. He/she has realised that TPM is operations-led and can, therefore, see only more work for him/herself and his/her staff. Some managers see TPM as a lifeline, others as a burden and it is the presenter's job to emphasise the benefits and that a TPM programme should be planned with the level of resources in mind.

We are undertaking a lot of other initiatives at present, people will see this as yet another burden that is imposed upon them.
This expresses what many of the executives may be feeling and recognises the 'flavour of the month' syndrome where one initiative after another seems to be imposed from above. The more enlightened executive will have realised that TPM can provide an extremely good 'umbrella' which will cover many of the existing initiatives and can be used to give them a clearer direction and meaning. Many improvement activities are carried out in a fragmented and uncoordinated way and the people involved in them find it difficult to understand what they are hoping to achieve.

The factory floor personnel and their unions will not like this. I don't think it will work here.
This comment is usually made because, first, the person is somewhat intimidated by factory floor personnel and in particular, union representatives and also has very little knowledge of the factory floor environment. Second, the person is not very enthusiastic about TPM because of the additional workload he/she perceives or the erosion of his/her 'empire' and is using this as a diversionary tactic. In most cases, this comment is untrue and is typical of the 'us and them' situation which exists within the typical scenario.

I have found that presenting actual examples and case studies can help the presentation considerably. Without the enthusiasm and commitment of the business executives then the TPM programme cannot even get off the ground. It is quite usual for the executives to give a partial commitment to the introduction of TPM at this stage, reserving judgement until a pilot has been completed.

Thereafter, awareness training sessions need to be provided for managers, engineers and supervisors. These are the people who will have to support TPM and release their production and maintenance time and resources to take part in the TPM teams, so it is very important that they understand the principles and practices of TPM and the role which they are expected to play. In my experience it is the line managers through to supervisor level who are the most difficult to convince that TPM will provide them with benefits and, consequently, they do not easily accept change. The reasons behind this probably stem from the fact that they are so used to being the 'middle men' who come under a great deal of pressure from senior managers and factory floor personnel alike and have to live with and try to implement the decisions made by their senior managers (which are often difficult and unpleasant) and operate within an environment of poor morale, lack of enthusiasm and co-operation and poor communications. They are at the 'sharp end' of the crisis management which is prevalent in many operating companies and do not know any other way of working. Most line managers are very sceptical about TPM and do not believe that the necessary time, money and resources will be provided to make it work. The principles and philosophy behind TPM are not usually challenged and are normally accepted as common sense. They are justifiably very cynical about any 'new initiatives' because they have seen so many come and go over the years, the majority of which have failed through lack of management commitment.

At an appropriate time, it is prudent to begin discussions with employee representatives such as trades unions, first at a site level and then by means of an awareness session and discussion with representatives who are local to the selected pilot area. The way in which TPM is received will depend upon the general state of industrial relations within the company. I have found that being an independent party has distinct advantages when running such awareness sessions. It is best to exclude managers from the session and encourage the representatives to talk freely and openly about their perceptions of the company, its management and the state of industrial relations. It is very important to communicate clearly what TPM is and is not and to point out the benefits that factory floor personnel will derive from the programme. The message will be greeted with a degree of scepticism, apprehension and, on occasions, with some hostility. I have a great deal of empathy with factory floor personnel who have in many cases, I believe, been subjected to very poor management who have made bad decisions and alienated the work force.

However persuasive the trainer, the scepticism and reservations of factory floor personnel will not be removed until TPM is seen to work in practice and some tangible results are observed. The objective of the communications sessions should be to gain acceptance of the principle that a TPM pilot or pilots will be allowed to proceed and that a review will be held at the end of the pilot(s) to discuss the experience and decide the way forward.

At this point it is also necessary to introduce the site maintenance personnel to the principles and philosophy of TPM so that they will not see the pilot area activities as encroaching upon their sphere of activities and can be made aware of the benefits they will gain as their role changes from a reactive to a pro-active one.

The communications exercise thus far has ensured that all of the company personnel who need to know about TPM before implementation begins have been made aware of the principles, practices, philosophy and implications of TPM namely, executives or senior managers, middle and line managers, factory floor representatives and maintenance personnel.

There may be a temptation at this point to broadcast the message of TPM to all other company personnel. This should be resisted for the reasons that were discussed in Chapter 2, section 2.3 and illustrated in Figure 2.7, i.e. that if people are told about a 'great new initiative' and then nothing happens for weeks or months or even years then its credibility is severely affected and it is forgotten. For this reason I usually advocate that communications sessions are held for operational personnel working within and serving the locality of the pilot area no more than a few weeks before the pilot is to be launched. The session will be very similar to that which was held for factory floor representatives and maintenance personnel with the emphasis on the objectives and implications of the pilot and, in particular, on the practical activities which will take place. The communications session will include factory floor personnel from the chosen pilot area in order to 'get them on board' and persuade them to approach the TPM launch with a willingness to 'give it a try'. This principle of communicating with personnel only just prior to the launch of TPM activities in their area ensures that something tangible starts to happen within their 'interest span' and allows TPM to get off to a good start.

The whole communications exercise is well worth the time and effort required to smooth the way of the initial TPM implementation and remove any industrial relations 'hurdles' which may otherwise be encountered.

Managers are sometimes so remote from factory floor personnel that they believe morale and industrial relations to be very good when just the opposite is true. To try to implement TPM in these circumstances without going through an appropriate communications exercise is very difficult indeed, if not impossible.

Getting started

As previously suggested, it is best to select one or more pilot areas as a vehicle for implementing TPM in the company. It is advisable that the pilot is launched very soon after the communications programme, as outlined, has been completed. The benefits of pilot TPM schemes are as follows:

- TPM can be tested in the company.
- TPM can be seen in practice.
- Tangible benefits of TPM can be shown.
- Lessons are learned for future area implementations.
- The implications of TPM implementation are understood.
- The next implementation steps are planned with greater confidence.
- The appropriate pace of TPM implementation is established.

The TPM team will consist of personnel who work within or are closely associated with a particular area of the company and the facilities therein.

I have worked with TPM teams of between two and eight people, the core of which consisted of personnel who operate and set up the facilities within their area. Other personnel with complementary skills and knowledge can be added to the core personnel and, in particular, it is desirable to have a maintenance technician or engineer on the team. It is also necessary to appoint or elect a team leader or co-ordinator. The constitution of the team, the area to be selected and the machinery to be used for the pilot, all has to be discussed and agreed by the managers and appropriate personnel. Where shift working is in operation, this adds another dimension to the discussion. Should the team span more than one shift? Should it just involve one shift? How can the TPM activities be communicated between shifts? These are all questions which need to be addressed. It is often necessary to arrive at a compromise, try it out during the pilot and learn valuable lessons which can be applied to subsequent area implementations.

The following sections 5.2 to 5.8 discuss the practical steps required to implement all of the components of TPM as illustrated in the 'TPM pie' in Figure 3.1.

5.2 Establishing the condition of facilities, restoring and maintaining them

The first step is to set up a TPM team which can identify with a particular area and the operation of facilities within that area. The team needs to be trained in the principles and practices of TPM with the emphasis on the practical activities in which they will be involved. In my experience it is not sufficient just to train the team and expect them to then carry out the activities by themselves. Most members of the team will have no experience of team working of this sort and, therefore, they will require a great deal of help and guidance during the initial stages of a TPM launch.

Having set up and trained the team, the first stage of autonomous maintenance is to establish the condition and the status of the facilities which have been selected. The team should be involved in selecting the facilities and given some guidance as to how much TPM work they should undertake. The choice of subject will depend upon the following:

- The type of facilities in the area.
- The complexity of the facilities.
- The importance of specific facilities (i.e. how critical they are).
- The number of people in the TPM team.
- Their knowledge of particular facilities.
- The history of the area and facilities therein.
- The future plans for facilities within the area.

When selecting machinery on the shop floor, for instance, if the machinery is large and complex it may be prudent to concentrate on particular sections of the machinery, whereas if the machinery is fairly small and simply constructed it may be appropriate to select several items for the TPM launch.

Having selected a particular facility or facilities such as machinery, the team must ascertain the present status of that machinery. There is a need to identify any faults and operating problems with the machinery and to understand just how well it normally performs. The initial reaction of operators who are closely involved with the machinery day after day is often one of 'I don't think that there is much wrong with this machinery, just one or two minor things'. This is often proved wrong, mainly because the operator is so used to the imperfections and everyday problems that he no longer perceives them as problems. It is only when the team begins to question everything that he begins to 'stand back' and take a fresh look at the machinery. The next section will deal with the types of faults and operating problems that are commonly encountered and provides some guidance concerning what to look out for.

A very significant activity that is carried out at this stage is that every fault and operating problem is listed and a 'tag' tied to the machinery as close to the fault/problem as is reasonably possible. I usually provide red tags which are clipped into clear plastic pockets where there is a danger that they will get dirty or oily. The significance of the tags is that:

- Everyone in the company can see that there is a fault/problem with the machinery.
- Instantly, the environment has started to change and there is a visual reminder that TPM has started.
- The tags represent the deterioration of the condition and operating performance of the machinery.
- The removal of tags represents a restoration of the machinery to its 'basic condition'.
- The highly sceptical members of the team have for once had the opportunity to discuss their problems and their difficulties and what's more, they have all been listed and tagged.

The completion of the tagging exercise provides the first level of the understanding of the state of the machinery concerned and can often prove to be a 'voyage of discovery' for the team members. I have also encountered a few line managers who have been stunned by the number of faults identified and profess to have had no idea that the machinery was in such a poor state.

Put a tag where there is a fault/problem

The number of faults and problems discovered will depend upon the type of facilities and how badly they have been allowed to deteriorate over the years. I have seen between less than 10 and more than 70 faults/problems found and tags attached, and always a proportion of them (can be as much as 30 per cent) are safety-related. Many of the faults and problems listed are relatively small and easily rectified and may be regarded as insignificant. The rule is to list and tag *all* faults and problems, no matter how small, as the TPM philosophy recognises that many small improvements lead to a large improvement in performance.

The tagging exercise will have started to encourage the team and 'thaw out' the most sceptical and unimpressed members of the team. The next step is to restore the machinery to its 'basic condition', i.e. ensure that all of the faults and problems listed are rectified and tags removed.

Restoration begins with cleaning, a thorough clean of the machinery and the adjacent area is required. Machinery is very often dirty, oily and dusty and even machinery which looks clean from the outside is often very dirty inside. It is important that the TPM team cleans the machinery and that it is not sub-contracted to another party. There are circumstances where machinery is very dirty and needs the bulk of the dirt to be removed by steam cleaning or other methods and, in this case, it is reasonable for other parties to undertake an initial clean to remove the bulk of the dirt. But even after this attention, machinery will still require more detailed cleaning. The reasons why the team members need to carry out the cleaning exercise are as follows:

Clean all parts of the machinery

- Cleaning is inspection. By removing covers, cleaning mechanisms and crawling all over the machinery, more faults are found that were not apparent during the tagging exercise. These can then be tagged and added to the list.
- Through cleaning, the team members get to know the machinery better and often discover parts that they did not realise existed.
- Through cleaning, the team begins to cultivate ownership and pride in its machinery.
- Through cleaning, the source of dirt, oil leaks and debris is discovered and added to the list of faults.

The cleaning exercise can often be the second 'voyage of discovery', and some quite significant discoveries can be made.

Senior managers often get involved during the cleaning activities and work with the team with their overalls on. This reinforces the management commitment to TPM, that cleaning is important and that something different is happening. I would strongly advise this, as most managers who join in actually find that it is an enjoyable experience.

To clean all parts of the machinery thoroughly may take more than one session, depending upon the physical size and complexity of the machinery and just how dirty it was. To fully restore the machinery to its 'basic condition' all of the faults and operating problems have to be eliminated. The TPM team has overall responsibility for the restoration work, and it is at this point that management commitment to TPM begins to be really tested. Many of the faults and problems can be rectified by the members of the team, and they should be encouraged to undertake all of the repairs for which they have the necessary skills and competence. In order to do this they require to be released from normal production duties for pre-arranged working sessions. The machinery must also be released from production so that the team members can carry out the work. Some expenditure on the parts and materials will be required to carry out the repairs. It is vital that management support is forthcoming because the team members are just becoming enthusiastic at this stage, having overcome their initial scepticism and reluctance to participate in TPM activities. This often puts line managers in a difficult position as they will be subjected to a great deal of pressure to maintain production and meet deadlines, yet the team is requesting that production is stopped for TPM working sessions. There will rarely be a convenient time to implement TPM within an operating company. It will not be easy and will invariably cause some 'pain'. This is true of most things in life that are worth having and it is, in this case, the only way in which a better way of working can be achieved. Usually, some compromise regarding the timing and duration of working sessions can be reached, even if it means working some unusual hours.

The TPM team will prioritise the faults and problems and aim to rectify those that are deemed to be most significant or safety-related first of all. Some of the work required to rectify faults and problems will require skills and competencies that cannot be found within the TPM team. In this case, the team will identify the resources required and where they can be obtained. Thereafter, if they are internal resources, then some negotiation with departmental managers who control the resource will be necessary. This is an area where some management support is required as team members often lack the confidence, contacts or level of authority to make such arrangements. If external resources are required, then contractors need to be contacted, quotations obtained and orders raised. Again, this is an area where management support is required, particularly during the early stages of TPM implementation.

As mentioned earlier, many of the faults and problems found will be fairly minor and easily rectified (in my experience 90 per cent plus), and through team working sessions and the use of some specialist resources they can be cleared in six to eight weeks. Those faults and problems that then remain will be those that may take much more time and effort and, probably, specialist resources to rectify. The way in which these are dealt with will be covered in sections 5.3 and 5.5.

Establishing the condition of facilities and then restoring them to a 'basic condition' is key to the implementation of TPM within the company. It is quite time-consuming, and there may be a temptation to skip this component of TPM or not to carry it out thoroughly and instead, to move on quickly to other components which seem more interesting. I strongly recommend that the restoration work is carried out thoroughly for a number of reasons, including the following:

It is much more difficult to draw up autonomous maintenance procedures for facilities that are still suffering from faults and operating problems.

> There is little incentive to start looking for improvement ideas and implementing them if the facilities are still not restored to a 'basic condition'.
>
> The restoration of facilities, by the TPM team is an important step towards building pride and ownership and morale.

There is also a temptation to paint machinery, racks, the factory floor, etc. and put up notice boards and slogans before any significant restoration work has been completed. This is not recommended because it can be perceived as making TPM a purely superficial exercise which is being undertaken for the sake of appearances only. My opinion on the subject is that machinery, racks, the floor, etc. can be painted and notice boards erected to publicise TPM, and anything else done to make the general area look good only after a significant amount of rectification work has been completed. It is 'the icing on the cake' and such actions cannot be accused of being superficial when there is a list of faults and problems which have been rectified, and 'before and after' examples to show the tangible improvements that the team has made.

Following the restoration of the facilities, measures need to be put in place to maintain the recently restored 'basic condition' and to ensure that they are not allowed to deteriorate again. During the initial investigation and later restoration work, the team will have identified certain activities that need to be carried out on a regular basis including:

- Cleaning of mechanisms, slides, surfaces, etc.
- Checking lubricant levels and topping up.
- Applying oil or grease to moving parts.
- Checking parts for tightness and tightening if necessary.
- Checking parts for signs of wear and deterioration.
- Checking and changing filters, etc.

Figure 5.2. illustrates these activities which are the basis of 'autonomous' or local, TPM team-based maintenance.

These activities can now be listed and developed into autonomous maintenance procedures. It is important that the TPM team is involved in drafting the procedures and that the machinery operators agree with them.

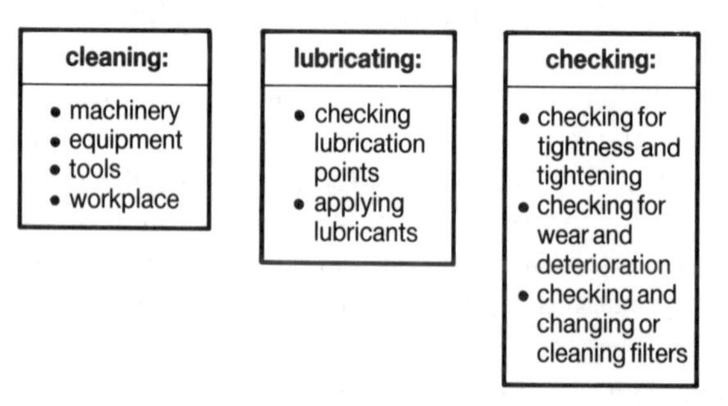

Figure 5.2 The basis of autonomous maintenance

The procedures should be simple and straightforward enough for anyone who is not used to operating or setting up the machinery to be able to understand them and carry them out without difficulty. The information contained in the procedures will include:

- What and where to clean, what to use, how often to clean and who is expected to do it.
- Which lubricant levels to check and where they are on the machinery, what lubricant to use if it needs topping up, how often to check it and who is expected to do it.
- What and where to apply oil and grease, how often to do it, what oil or grease to use and who is expected to do it.
- Which parts to check for tightness, where they are on the machinery, what to use, how often to check them and who is expected to carry out the check.
- Which parts to check for wear, where they are on the machinery, what to look for, any tools that are needed, how often to check them and who is expected to do it.
- What other checks, etc. to carry out, how often and who is expected to do them.

Some blank procedure sheets are provided in Appendix 6 along with examples of autonomous maintenance procedures.

Once the procedures have been compiled and agreed, they should be protected in some way and kept on or near to the machinery as a readily available reference. The activities listed on the procedures now have to be carried out at the intervals stated. Time has to be allowed for the operator, maintenance craftsman, etc. to comply with the procedures, A ticklist should also be provided which can be marked off when each activity is completed, and the procedures can be audited from time to time by the TPM team to ensure that they are effective, relevant and practical and that they are being carried out. If necessary, they can be revised and/or actions taken by the team to ensure that they are carried out.

As a result of regularly carrying out the autonomous maintenance procedures, the general condition of facilities will be maintained, the number of breakdowns or stoppages will be substantially reduced as will inconsistent quality and operation.

5.3 Identifying and eliminating faults and operating problems

In Chapter 3, section 3.3 the difference between faults and operating problems was explained and the different types of faults and problems were discussed. The inspection of facilities is an important part of the early stage of TPM implementation and is the first step to autonomous maintenance. Let us now identify the types of faults and operating problems that are commonly encountered, and the actions that are necessary to eliminate them.

Faults commonly encountered are namely:

Machinery or equipment structure:

- covers missing or not fitted correctly;
  Action: replace or refit covers
- screws, nuts or bolts missing from covers;
  Action: replace
- seals missing or worn on slideways;
  Action: replace or refit seals
- too much or too little lubrication on moving parts;
  Action: investigate reason and clear any blockage and adjust
- oil drips from slideways, etc.;
  Action: adjust, investigate return route, clear any blockage and fit drip tray if necessary
- slides are worn and cause excessive 'play';
  Action: probably requires refurbishment/replacement of worn parts
- screws missing from guards;
  Action: replace screws
- guard sections missing or not fitted correctly;
  Action: investigate where they have gone and refit or replace
- guard interlock not working or not fitted;
  Action: investigate, replace switches/sensors if necessary, fit interlock if necessary
- machinery not secured to floor;
  Action: move machinery, drill floor and fit bolts, re-locate and bolt down – may need outside contractor
- services not properly piped/wired to machinery or equipment;
  Action: re-route pipes/wires if necessary, secure
- inspection windows dirty or cracked or frosted;
  Action: replace, investigate better material
- door seals worn or incorrectly fitted;
  Action: replace seals

Drive elements:

- lubricant site level dirty and difficult to read;
  Action: drain lubricant, clean and/or replace or re-position site level. If not possible use a mirror mounted on a stick.
- low lubricant level;
  Action: top up with correct lubricant
- drive belt or chain tension incorrect;
  Action: adjust the tension
- drive belt worn or frayed;
  Action: replace belt and set tension

- drive belt covered in contaminant;
  Action: investigate source of contamination and eliminate if possible, fit cover around belt
- drive pulley/sprocket broken;
  Action: replace
- oil leak from gearbox;
  Action: investigate exact site after cleaning, replace or refit seal or gasket
- clutch worn and malfunctions;
  Action: take apart, check, replace clutch or parts and re-assemble
- brakes not working properly;
  Action: take apart, investigate, replace worn parts and refit
- bearings worn – excessive play;
  Action: remove bearings and check, replace bearings and shafts if necessary
- electric motor end cover missing or broken;
  Action: replace if possible, if not, mend with adhesive
- motor cooling fan broken or catching on cover;
  Action: investigate and reposition or replace the fan

Pneumatics and hydraulics:

- filter is old and clogged up;
  Action: replace or at least clean thoroughly
- pressure gauge not working correctly;
  Action: replace
- pressure gauge glass broken;
  Action: replace
- valve leaking;
  Action: investigate, tighten fittings, replace seals or replace valve
- screws missing from valve mounting;
  Action: replace screws
- valve not mounted correctly;
  Action: reposition and refit
- pipes or fittings leaking;
  Action: replace pipes, investigate using better material, tighten fitting or replace
- flexible pipes worn or frayed;
  Action: replace, investigate reason for fraying and try to eliminate
- incorrect pipe length;
  Action: shorten or replace with longer
- cylinder mounting worn;
  Action: replace mounting or remove bush and replace
- screws missing from mounting or loose mounting;
  Action: replace screws, tighten
- site level on reservoir dirty and not easy to read;
  Action: drain fluid, clean and replace or reposition if necessary

Electrical equipment:

- wire insulation frayed or cracked;
  Action: replace
- connections loose;
  Action: tighten or replace if necessary
- incorrect wire or trunking used;
  Action: replace with correct standard
- wires, conduit or trunking not secured correctly;
  Action: secure with correct clips or mountings
- screws missing from connection box cover or mounting;
  Action: replace screws
- limit switch dirty and sticking;
  Action: clean and test, replace if necessary
- limit switch mounting loose;
  Action: tighten mounting screws
- limit switch mounting inadequate;
  Action: replace with more robust mounting bracket or plate
- rubber seal on switch stem worn, cracked or missing;
  Action: replace
- switch or sensor unprotected from dirt or debris;
  Action: reposition and/or fit a cover
- seal on electrical cabinet door worn or missing;
  Action: replace
- door on electrical cabinet not secure;
  Action: fit or repair lock
- holes in cabinet – dirt, oil or debris inside;
  Action: fill in holes
- push button broken;
  Action: replace
- push button not working correctly;
  Action: investigate and replace if necessary
- lamp or lens broken;
  Action: replace
- contactor sticking on or off;
  Action: investigate, clean, replace if necessary
- solenoid sticking on or off;
  Action: investigate, replace if necessary
- sensor too coarse or too fine;
  Action: investigate range of operation and replace or adjust
- sensor not robust enough – keeps failing;
  Action: investigate alternative and replace
- software 'bug' in control programme;
  Action: investigate and eliminate

Operating problems commonly encountered include the following:

Machinery and equipment structure:

- access door very stiff and heavy;
  Action: investigate reason, clean or replace rollers, replace door mechanism
- guard restricts operation;
  Action: look at the method of operation and guard design, re-design guard and replace
- inadequate guard – unsafe and/or splashing;
  Action: re-design guard and replace
- debris, oil, fluid sprays on to operator;
  Action: investigate source and design cover/deflector to eliminate the spray
- debris, oil, fluid sprays on to the floor or structure;
  Action: investigate source and design cover/deflector or catcher to eliminate the spray or catch it before it goes on the floor or structure
- holes or pockets in the machinery structure;
  Action: fill in or cover holes up
- holes in structure allow fluids to mix;
  Action: fill in holes
- process waste is difficult to remove;
  Action: investigate improved methods or design changes and implement
- insufficient lighting;
  Action: fit additional lights
- access for setting and adjustment is difficult
  Action: investigate and re-design the machine access or tooling if necessary
- process fluid spray control is inadequate;
  Action: fit improved nozzles and valves
- process waste builds up in process area;
  Action: investigate methods of removing waste through re-design of the structure, fitting conveyor, etc. and implement

Drive elements:

- oil site level difficult to see and access;
  Action: reposition site level
- filler cap difficult to access;
  Action: reposition filler cap or adjacent equipment to make access better. If not possible, design special filler arrangement
- oil drain difficult to access;
  Action: reposition drain or adjacent equipment if possible. If not, design and fit a special drain arrangement
- no tension adjustment on belt or chain drive;
  Action: design and fit adjuster if necessary

- grease nipples missing or difficult to access;
  Action: refit missing nipples, reposition if possible or obtain a special fitting for grease gun
- motor/gearbox removal difficult;
  Action: if deemed necessary, re-design structure or reposition adjacent equipment to improve access

Hydraulics and pneumatics:

- reservoir site level difficult to see and access;
  Action: reposition site level or if not possible, provide mirror on a stick
- filler cap difficult to access;
  Action: reposition filler cap or adjacent equipment to make access better. If not possible, design special filler arrangement
- drain plug difficult to access;
  Action: reposition drain or adjacent equipment if possible. If not, design and fit a special drain arrangement
- pipe runs across access area – tripping hazard;
  Action: re-route pipes, lift up or sink into floor
- pipe runs impede covers and guard removal;
  Action: re-route pipes
- pipe runs form pockets for dirt, debris and oil to collect;
  Action: re-route pipes or fill in pockets
- pressure gauges are difficult to see and access;
  Action: reposition
- operator unsure of correct pressure;
  Action: mark up gauges with red and green showing normal working pressure range and alarm range
- pressure valve adjustment not locked in position – operator unsure of correct position;
  Action: mark up valve handle/body with correct position and lock valve if possible
- pipes and fluid flow difficult to identify;
  Action: identify all pipes and direction of flow

Electrical equipment:

- wiring untidy – tripping or catching hazard;
  Action: re-route and tidy up wiring
- wire runs impede covers or guards;
  Action: re-route wire runs
- wire runs cause pockets for dirt, debris and oil to accumulate;
  Action: re-route wire runs or fill in pockets
- legend missing from control panels or badly worn;
  Action: replace
- readout remote and hard to see;
  Action: reposition

- wires and terminations not clearly identified – difficult to find fault;
  Action: fit indicators to terminations and wires
- push buttons and dials inconveniently positioned for operation or setting;
  Action: reposition
- insufficient warning of machinery faults;
  Action: investigate condition monitoring techniques and fit the appropriate sensors
- inadequate diagnostics for identifying faults;
  Action: improve the documentation

General:

- documentation very poor, difficult to read and understand, no simple work instructions for machinery operation, care and maintenance;
  Action: produce autonomous maintenance procedures and standard working sheets. Obtain a full set of supplier's documentation
- working environment poor, dirty, difficult and a little dangerous;
  Action: implement TPM
- inadequate training provided for operation, set up and maintenance;
  Action: implement a comprehensive training programme

The reader can use the list as a guide when undertaking the inspection of facilities and, although the list is biased towards machinery faults and operating problems, many of the items apply to any equipment and work stations.

5.4 Measuring the effectiveness of facilities

With any improvement activities it is necessary to put in place some relevant measures of performance which serve two main purposes; first, to measure the effect of the improvements and thus direct improvement activity, and second, to enable the financial return on the activities to be calculated.

In Chapter 3, the reader was introduced to the concept of the six big losses and through a number of examples, shown how they could be measured and converted into a figure for the overall effectiveness of the facilities. How, then, in practice, can information be gathered pertaining to the losses and overall effectiveness calculated and used? In general, some of the information which is required is usually available. Many operating companies record information such as the following:

- The number of units produced in a given period.
- The number of rejects produced.
- The number of hours worked during a particular period.
- The amount of down-time due to breakdowns.

In addition, some may gather information about change-over and set-up times and the number of units that have to be re-processed. Quite often, however, the information is gathered by different individuals or departments such as quality, production control or finance.

The initial stage in measuring the effectiveness of facilities is to recognise the six big losses in that particular environment and, therefore, what information will need to be gathered in order to measure the losses. In the examples shown in Chapter 3, it can be seen that the kind of losses differ from high-volume production using machinery to low-volume 'hand' assembly. In the former case, for instance, when measuring availability, it is the time that the machinery is broken down and the time taken to change over tooling and fixtures which is recorded, whereas in the latter case, availability is a product of the time lost due to the lack of availability of facilities or parts and the time taken to adapt facilities for that particular assembly. So, although the principles of measuring the six big losses are the same, the definition of them and method of measurement will vary.

The TPM team, in conjunction with the personnel who presently gather pertinent information, will need to establish the information required and then ask 'how can we collect this information on a regular basis?' In many cases it is appropriate that the personnel should gather the information on the losses.

Appendix 5 contains a typical effectiveness information collection sheet. In this case the information is gathered as follows:

- The area, machine and date is written at the top of the sheet.
- The shift date and time are entered in the first column.

Operators are often the best people to gather information

- The job reference is entered in the second column (optional).
- The time that the job was started and stopped is written in the third column and the total number of loading hours recorded (depending upon the definition of loading time, the time taken by events such as breaks and meetings may be deducted).
- The amount of time taken to change over and set up the job is entered in the fourth column.
- During the loading time, any hours lost due to breakdowns are recorded in the fifth column.
- The number of units actually produced is entered in the next column alongside which is the standard production rate for the unit.
- In the last two columns, the number of scrapped or re-processed units is recorded.

The following sheet in Appendix 5 shows an example of a sheet which has been filled in for one week and how the total loading time, total numbers produced of each type and total scrap and re-work figures are compiled at the bottom of the sheet.

For low-volume applications where it is not possible to gather meaningful information pertaining to the numbers produced and numbers scrapped, the figures in the performance and quality columns will record the amount of time lost due to minor stoppages or slow speed operation and time lost due to poor quality.

The information gathered can now be used to calculate percentage figures for availability, performance and quality and overall effectiveness. Initially, the team members will need some help to carry out the calculation and may feel it necessay to design their own information collection sheet. As they become more conversant with the calculations, they will be able to calculate the effectiveness of their own facilities. It is important that the information gathered is used and is seen to be used in a positive way. The percentage figures should be plotted on a chart which will be displayed in close proximity to the facilities. This will ensure that the overall effectiveness figures are very visible. Figure 5.3 shows the layout for an overall effectiveness chart and dimensions are provided in Appendix 5.

There is a temptation to use the overall effectiveness figures as a means of comparing facilities, areas and operating companies. This should only be undertaken with great care as any differences in the calculations and methods of gathering data will make the comparison difficult and may de-motivate people. The main use of overall effectiveness figures should be:

- As a means of monitoring the status of the facilities – are they good or bad, getting better or worse?
- To direct improvement activity – which losses limit the effectiveness figure and how significant are they?

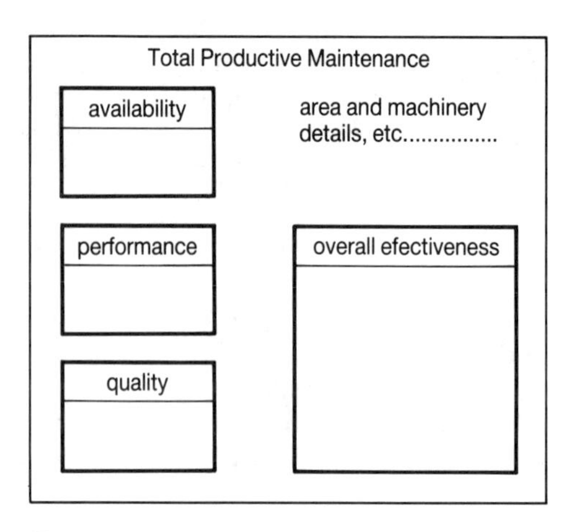

Figure 5.3 An overall effectiveness chart

■ 5.5 Establishing a clean and tidy workplace

The disciplines of 'CAN DO' were explained in Chapter 3 and are very much associated with the general work area. This is an important aspect of TPM implementation and one which makes the biggest visual impact. Sections 5.2 and 5.3 concentrated on restoring facilities and implementing autonomous maintenance which is specifically focused on the facilities within the area and, in particular, machinery and equipment. Cleanliness of the facilities has already been described as an important part of restoration and cleaning will, therefore, be undertaken during the team working sessions. The application of 'CAN DO' disciplines is very complementary to autonomous maintenance activities and encourages the TPM team to look at the workplace and all other facilities.

My approach when launching a TPM team is to encourage them to look at the major facilities first, but also to address the general work area and ask 'which facilities do I need to carry out my job?' This leads to:

A list of all of the facilities used, right down to a small screwdriver, being produced.

The availability of those facilities being questioned.

The location of the facilities being investigated.

By undertaking this exercise, any deficiencies are identified and also any rubbish or redundant facilities can be removed and disposed of. Often, some small quantities of hand tools, fixtures, racks, etc. have to be purchased or manufactured because personnel either do not have them or do not have enough of them. A lot of time is wasted in operating companies through

Throw away anything you don't need

personnel having to search for the correct item or having to make do with the incorrect tool for the job, all for the cost of a few pounds!

The disposal of rubbish and facilities that are no longer used releases more space and quite often enables facilities to be repositioned for ease of use. If there is less clutter about, it is easier and more incentive is provided to keep the general work area clean and tidy. Neatness is achieved by having a defined place for everything, and the team should endeavour to find a location for all of the tools, fixtures, materials, etc. that are used to do the job. Cabinets, racks, tables and boards can be used to store items, but it is desirable to identify which tool or material should be at which location. This can be done by listing the tools which should be at a certain location, labelling the location or, best of all, shadowing the tool on a board or on a shelf. This is achieved by painting the outline of the tool on the board or shelf and perhaps writing the name or reference of the tool within the outline. This has the advantage of providing a visual guide as to where tools should be located and instant indication when a tool is missing. Figure 5.4 shows a typical shadow board for hand tools. The discipline of always returning the tool, fixture, material, etc. to its proper location will gradually be established, once a place for everything has been provided.

Where portable facilities are used which are regularly moved to where they are required to do a particular job, the same principle can be applied and a defined location provided, identified and marked on the floor. This ensures that personnel will not have to spend valuable time looking for facilities which could be anywhere in the area. To this end, it can help if such facilities are

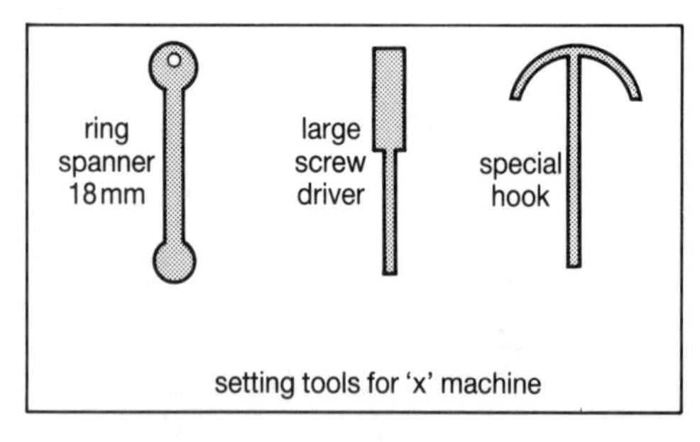

Figure 5.4 A shadow board

painted a bright colour, such that they stand out from the other facilities and are easily spotted if they are not at their defined location.

The implementation of 'CAN DO' disciplines is a gradual and ongoing process of change which will be achieved by many small improvements to the working area. Once the workplace is tidier and cleaner, it is appropriate for cabinets, tables, racks, floors, structural steelwork, etc. to be painted and for TPM information to be displayed on notice boards. This has quite a big visual impact and is the mark of a TPM company, provided it can be backed up by evidence of the thorough cleaning, restoring and improvement of facilities.

■ 5.6 Identifying and eliminating significant faults

As discussed in Chapter 3, the autonomous maintenance activities will naturally reveal any significant faults which are inherent in the design of machinery, fixtures, tooling, etc. or in the methods and systems employed to carry out the process. The analysis of overall effectiveness figures will highlight major problem areas and the effect that inherent faults have on the performance of the area. The fault that has been identified has been shown to be significant, and in many cases it will be quite complex and will require a great deal of time, effort and expertise to resolve; therefore, it will need to be approached in a way which is different from normal autonomous maintenance. First, the project will need to be cost-justified on the basis of an improvement in overall effectiveness or some other measure of performance. In many cases, the projected amount of improvement in overall effectiveness can be stated in financial terms and used to justify the investment in the project. The following example can be used to illustrate this:

In Chapter 3, section 3.4, example 3 described a fault which often occurred because two parts were not assembled correctly (no. 1) and this resulted in a minor stoppage and the loss of one unit. The frequency of stoppages encountered made a significant contribution to the low level of performance (70 per cent) and overall effectiveness (42 per cent). The TPM team decided that these losses were not acceptable, and initial investigation showed that a re-design of the machinery fixtures was required in order to eliminate the fault. Some simple ideas were tried, but were not successful and it became clear that a focused improvement project was necessary. The cost justification for this could be based upon the overall effectiveness figure which the team estimated would be achieved once the fault was eliminated. Without these losses the overall effectiveness could be recalculated as follows:

% availability would remain as 73%

new performance losses = $(15 \times 5) + (5 \times 2 \times 5) = 125$ mins/wk

$$\% \text{ performance} = \frac{1650 - 125}{1650} \times 100 = \frac{1525}{1650} \times 100$$

$= 92\%$ (an improvement of 22%)

new quality losses = $(30 \times 1) + (220 \times 1) + (24 \times 3) = 322$ units/wk

$$\% \text{ quality} = \frac{2875 - 322}{2875} \times 100 = \frac{2553}{2875} \times 100$$

$= 89\%$ (an improvement of 7%)

overall machine effectiveness = $0.73 \times 0.92 \times 0.89 \times 100$

OME = 60% (an improvement of 18%)

Based upon this projected improvement in overall effectiveness, the cost savings could be shown thus – the additional earning capacity for each 1 per cent improvement in overall effectiveness was calculated in Chapter 4, section 4.6 as £405. Therefore, the savings generated =

£405/1%/yr × 18% = £7290 per year

The team estimated the cost of the project as follows:

- team man hours = 4 people × 20 hours × 10/hr = £800
- engineering designer man hours = 40 hours × £15 = £600
- cost of purchasing new fixtures = £3600
- maintenance technician man hours = 20 hours × £15 (premium time) = £300

No production loss would be incurred if the work was carried out during a weekend. The total estimated project cost = £5300

By comparing the estimated project cost with the projected savings, a financial payback for the project could be calculated as follows:

$$\text{Payback} = \frac{£5300}{£7290} \times 12 = \text{less than 9 months}$$

Most operating companies would accept this level of payback.

This method of cost justification can be applied to all projects which will have a direct bearing on the overall effectiveness of facilities, provided that the existing figures are available and realistic estimates for the projected level of overall effectiveness and project costs can be compiled.

Project cost justification can also be based upon improvement of the utilisation of facilities, where the loading time is increased due to improved scheduling and organisation of manufacturing resources. This would be relevant for focused improvement projects which are related to loading facilities more effectively, rather than improving the overall effectiveness of them.

Having identified the need for a focused improvement project, analysed its costs and benefits and proved that it is financially justified, the next step is to set up the project. Setting up the project will often involve personnel other than the TPM team and most probably, the site TPM co-ordinator will have to be called upon for assistance. The skills and resources that are necessary to carry out the project have to be identified and usually, because of the more complex nature of focused projects, people external to the TPM team will be required.

In the example previously discussed, it is clear that specialist design and manufacturing resources will be needed in addition to the maintenance technician who will fit the new fixtures. Typically, for such a project, the team would consist of:

- The operator of the machinery (who will probably be a member of the TPM team).
- The craftsman who usually sets the machinery up (who also will probably be a member of the TPM team).
- An engineering designer, with a knowledge of fixture design for this type of machinery (either employed by the company or sub-contract).
- A maintenance technician with the necessary fitting skills.

The team may call upon other resources within the company and/or external suppliers as necessary, but would have overall responsibility for the implementation of the project.

Carrying out focused improvement projects

The key to the successful implementation of focused projects is that they should be approached in a structured way, i.e. undertaken in a logical sequence, step by step. Some guidelines to the implementation of different kinds of focused project are provided as follows:

Availability improvement projects – changeover time reduction

There are three major aspects of changeover from part to part which need to be considered in this kind of project:

1. The methods that are used to carry out the changeover, including all of the facilities used.
2. The design of tooling and how it is assembled to the machinery.

3. How the changes from part to part are scheduled.

The steps are as follows:

1. Identify the tooling and parts which give the significant problems.
2. Study a typical changeover.
3. List each element of changeover.
4. Time each element of changeover.
5. Identify which elements of changeover are 'internal' (take place on the machinery, whilst it is not producing).
6. Identify which elements of changeover are 'external' (take place off the machine and do not stop it producing).
7. Question each element of changeover – can it be eliminated? How can the time be reduced? Can 'internal' elements be converted to 'external' elements? Consider the different options.
8. Decide upon the best solutions and identify action that is necessary.
9. Plan timescales and resources to implement the action.
10. Implement action and measure results.

The diagram in Figure 5.5 illustrates the main steps.

It is quite usual to halve the changeover time mostly as a result of improved methods and some small hardware improvements. The project will bring about a substantial improvement in changeover time and thus the availability of the machinery. Further continuous improvement activity should thereafter be used to keep gradually reducing changeover times and improving flexibility as illustrated in Figure 5.6.

Availability improvement projects – reliability improvement

Where machinery is frequently breaking down due to a clearly identified cause, then the reliability improvement project team can begin to brainstorm solutions almost immediately, assess them and decide upon the most cost-effective

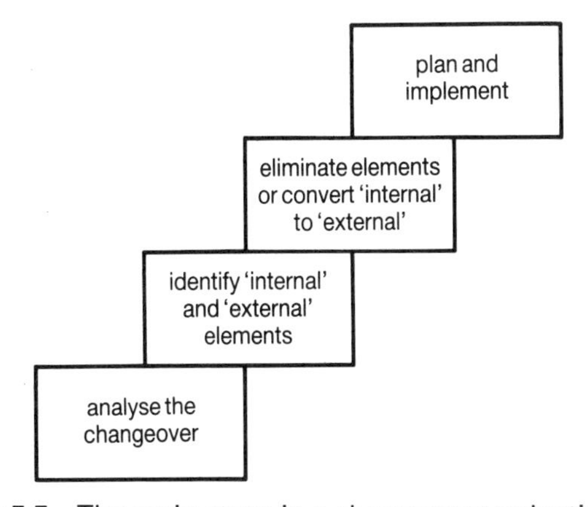

Figure 5.5 The main steps in a changeover reduction project

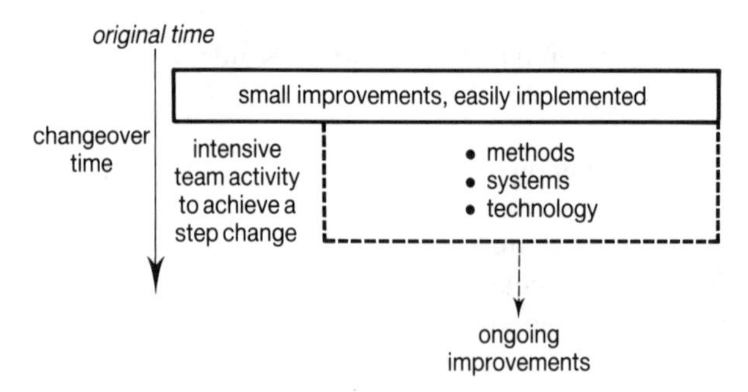

Figure 5.6 The pattern of changeover improvements

solution. On many occasions, however, the situation is not so straightforward and the source of poor reliability is not evident. Under these circumstances the steps below should be followed:

1. Analyse the historical information on the machinery including the number, type and frequency of breakdowns and amount of time spent repairing the machinery.
2. If insufficient information is available, arrange for information to be gathered over a time period which will provide a typical set of results under normal operating conditions.
3. Analyse the information: categorise the types of faults and problems encountered along with where they occur on the machinery or during the process and present them on a histogram.
4. Discuss the findings and agree the priority areas to be addressed.
5. Analyse the faults using a structured problem-solving technique. This will be explained later in this chapter.
6. Brainstorm solutions for the highest priority areas first (the principles of brainstorming are detailed later in this chapter).
7. Decide upon the most cost-effective solutions.
8. Plan and implement the solutions.
9. Move on to the next area and repeat steps 5 to 8, whilst still monitoring the performance of the machinery and the effects of the action so far.

It is quite usual to find that there are just a few areas and/or recurring faults which cause the majority of breakdowns and, by addressing these first, quite rapid benefits are achieved.

There are a number of very good problem-solving techniques which can be used by the project team to analyse the fault and establish its root cause. Two of these techniques are described below, along with the brainstorming technique which is used primarily to generate ideas:

1. Cause and effects analysis (fishbone diagrams)

This is a very useful technique which is best carried out by the team using a flip chart or white board. The exercise begins with a definition of the effect of the fault or operating problem. For example, this may be 'the cutting head does not operate and thus the machinery cannot function correctly'. This effect is then written on the board alongside a bold, horizontal arrow (usually pointing left to right). The team then start exploring what could be the primary causes of the fault. For example, these could be 'the motor is not operating, or the gearbox is not transmitting the drive, or the clutch is not operating or the electrical controls are not working'.

These primary causes are then written on the board alongside arrows which point directly at the horizontal arrow. Each of the primary causes is then analysed by the team and the secondary causes established. For example, for 'the motor is not operating' the causes could be 'there is no power applied to the motor, or its coils are burned out, or it is mechanically seized, or it has an internal electrical fault'. The secondary causes are then written down alongside arrows which point directly to the appropriate primary cause arrow. The cause and effects diagram is gradually built up and a picture of the possible causes of the fault or problem emerges. The team pursues each cause as far back to its root as much as is deemed necessary to establish the most likely root cause of the fault or problem. Figure 5.7 shows the structure of a cause and effects diagram.

2. The five whys

This is a similar technique for fault analysis where the individual or team is encouraged to ask 'why' five times when faced with a problem or fault in order to establish the root cause or causes. The technique questions everything about the fault and does not take any explanation at face value. For example, 'the limit switch is always malfunctioning and causing the equipment cycle to stop', WHY? – 'because it keeps sticking', WHY? – 'because it gets very dirty', WHY? – 'because dirt and oil drip on to it', WHY? – 'because there is nothing to stop the dirt and oil dripping on to it', WHY? – 'because it does not have a cover and there is an oil leak just above it'.

From the example, what began as the description of a particular fault has developed into a statement on the root cause of the fault, a solution for which can either be to solve the oil leak or fit a cover to the switch or perhaps to do both.

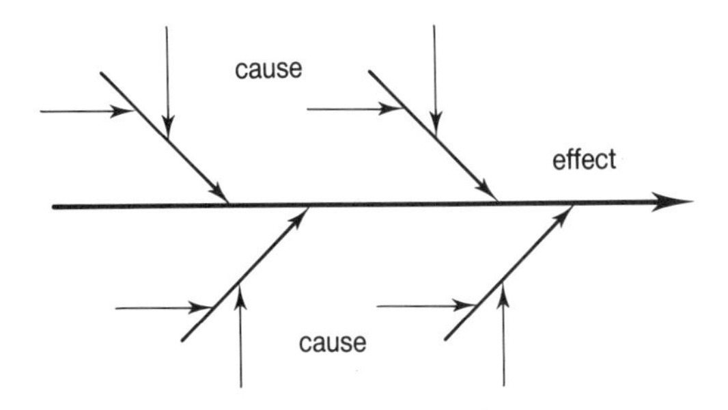

Figure 5.7 Cause and effects analysis

3. Brainstorming

Brainstorming is a team-based technique for generating ideas and ensuring that all possible options are considered. The rules of brainstorming are:

- Use a flip chart and get one of the team to write all of the ideas down on the flip chart.
- Clearly state the problem and the objective of the exercise.
- Go around each member of the team in turn and ask them to generate a solution to the problem.
- At this stage, any suggestions must be written down, no matter how outrageous, and none should be questioned.
- Keep going around the team until no more suggestions are forthcoming.
- The team assesses each idea in turn and judges it against the requirements and constraints.
- The team agrees those ideas which are not feasible, do not meet the requirements or are not possible within the stated constraints. They are crossed out.
- The remaining ideas are discussed in more detail and assessed again until a preferred option is agreed.

Performance improvement projects – reducing cycle time

Where the throughput of the process is significantly below what is expected, the project team will initially need to establish the possible causes of the problem. To accomplish this the early stages are:

1. List each activity that takes place during the operating cycle. This should include all machinery actuations, processing, operator activities and delays and can be carried out at a very detailed level if necessary.
2. Where possible, time each activity or major group of activities with a stopwatch. Some machinery control systems can be programmed to time each activity and print out the results. The timing exercise should be carried out a number of times, at different times of the day and for different operators.
3. Analyse the activities and their times, looking for trends and variations and also identify the activities which take up the most significant amount of the cycle time.
4. Based upon this analysis, the particularly significant problems which are besetting the operation of the machinery can be identified.
5. Establish their root cause using one of the structured problem-solving techniques previously described.
6. Brainstorm the possible options and decide upon the most cost-effective solution.
7. Plan and implement the improvement.

Quality improvement projects – mistake proofing processes

The prevention of defects at source is central to the philosophy of many world-class manufacturing businesses, and 'poka yoke' or mistake proofing is a proven approach to the elimination of many process quality problems. Poka yoke is particularly relevant to production processes which involve a fair amount of operator interaction, and its aim is to ensure that things cannot go wrong. The improvement project team starts by identifying the workstations or stages of the process where problems most often occur, along with the type of problems encountered. They then ask the question 'how can we ensure that the problem cannot happen?', brainstorm ideas and decide if any of them are feasible. An example is where two parts can be placed into a fixture the wrong way round. This sometimes happens and results in defects. In this case, the team would investigate ways of making sure that the parts cannot physically be located in the fixture the wrong way round. They may require the fixture to be re-designed. If no solution of this type can be found, then the team asks the question 'If the error occurs then what can we do to ensure that it is detected straightaway, before a defect is produced?' In the example, if it was not possible or not viable to re-design the fixture then perhaps sensors could be fitted around the fixture which would indicate whether the assembly was good or bad before the process continued. There are many types of poka yoke device with three main actions. The main action of a poka yoke device is described as follows:

1. Physical prevention of an error from happening such as restraints on a fixture which prevent the parts being incorrectly loaded (the pin configuration on many electrical plugs serves this same purpose).
2. Detection that an error has occurred and indication that it has happened such as sensors that check the orientation of parts and cause an audible or visual alarm to be actuated.
3. Detection that an error has occurred and prevention of the process from proceeding such as sensors that check the orientation of parts and send a signal to the machinery control system which inhibits its operation.

Types of poka yoke device can implement the following procedures:

1. Check that an operation has been carried out correctly.
2. Check that the correct number of operations has been completed.
3. Check that the correct sequence has been completed.

The team will need to carry out the analysis of the situation and having agreed an option, design the poka yoke device or system to suit the application.

Team training

Focused improvement project teams will need to receive training in the techniques discussed and also in the structured approach to problem-solving and implementation. It is not reasonable to expect a project to be successfully completed and the projected benefits to be achieved if the team is not properly trained. Such training is an investment as very similar techniques and approaches are used for different types of improvement project and, once personnel have been trained and have used these techniques in practice, they will be equipped to contribute to future projects. The training and experience of being involved in a focused improvement project is also an effective means of developing personnel and realising their potential.

■ 5.7 Providing maintenance systems to support facilities

In Chapter 3, section 3.2 the relative roles of operations and maintenance personnel were discussed and these were illustrated in Figure 3.3. The activities undertaken by operations personnel, through autonomous maintenance, and their effect upon the general condition and performance of facilities has been described along with the new role required of maintenance personnel. Their function becomes much more pro-active, always seeking to prevent breakdowns, to anticipate performance and quality problems rather than rectify them after they have happened. To fulfil this role in an effective and professional manner, maintenance personnel need new systems which will support their activities. Unlike many of the components of TPM which have been discussed so far, this is very much the responsibility of maintenance personnel and they must lead the development and implementation of new systems and working methods.

It is acknowledged that some operating companies do not have a maintenance department or some areas of the company are not served by maintenance personnel. In these cases, systems are still required for the provision of specialist service personnel, whenever the need arises.

The recommended steps to the implementation of maintenance systems are as follows:

Audit the existing systems and methods of operation of the maintenance department (or other parties). It is necessary to understand the present position, strengths and weaknesses. The audit can be carried out by a team of maintenance personnel, but it is usually much more objective if the audit is provided by an external party.

Assemble a small team of relevant maintenance personnel and managers, as appropriate, and discuss the findings of the audit along with the future role of maintenance personnel.

Agree and compile a maintenance strategy for the site which will give a clear indication of the direction in which the company is heading and the overall objectives for maintenance personnel.

Agree relevant measures of performance for the department and ensure that they are directly related to the overall effectiveness of machinery and a pro-active approach.

Decide what is required from the maintenance systems and define their scope.

Develop or adapt existing manual-based systems to support both the TPM teams which are carrying out autonomous maintenance activities and maintenance personnel who are responsible for the following:

- responding to breakdowns and carrying out repairs;
- replacing worn or contaminated parts;
- carrying out regular planned maintenance (particularly those activities which have been identified on the autonomous maintenance procedures);
- implementing focused improvement projects as required;
- recording information on the fault history of machinery, the cost of repairs, etc.;
- investigating the use of predictive maintenance techniques, cost justifying, implementing and operating them;
- controlling the storage, usage and purchase of spare parts.

Implement the systems in stages and continuously seek to improve and develop every aspect of the system.

If appropriate, specify and implement a computerised maintenance management system gradually to replace the manual systems and improve the effectiveness of the system.

Figure 5.8 illustrates the main steps to the achievement of effective maintenance systems.

At each stage of the evolution of maintenance systems it will be necessary to train and develop maintenance personnel. Initially, they will require an understanding of the principles, practices and philosophy of TPM, followed by training related to maintenance techniques and systems.

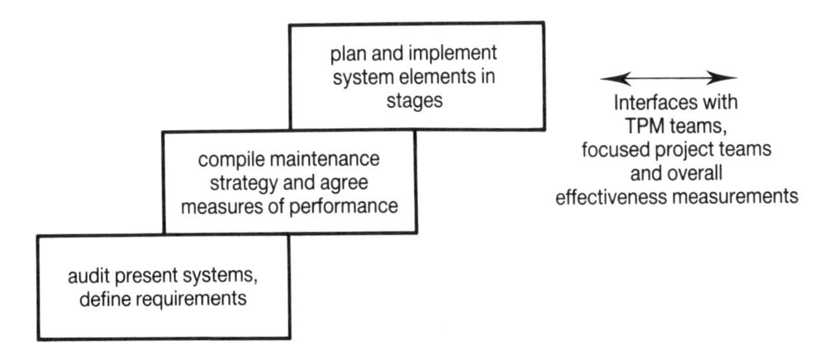

Figure 5.8 The steps to maintenance systems implementation

■ 5.8 Purchasing and installing facilities that provide the best return

The components of TPM that have been detailed so far in this chapter have concerned the implementation of basic TPM disciplines in the workplace, i.e. good housekeeping, autonomous maintenance, measuring and improving overall effectiveness and maintenance systems which have all been orientated around facilities that already exist within the company.

One other important component of TPM is necessary – ensuring that any new facilities are purchased with the TPM philosophy and detailed lessons learned from operating existing facilities very much in mind. The Japanese often refer to this component of TPM as 'early equipment management' which deals with the planning, specification, design and installation of new facilities. For the purposes of this book, I will refer to this component of TPM as 'machinery management'. The main requirements for machinery management are as follows:

- Development – the continuous development of improved production methods and processes.
- Reliability – specifying and designing machinery which is inherently reliable.
- Economy – achieving a balance between machinery performance, capital cost and operating costs (life-cycle costs).
- Availability – achieving high levels of machinery availability for production with the minimum of down-time.
- Maintainability – specifying and designing machinery which needs the minimum of maintenance makes maintenance tasks and fault diagnosis easy.

The implementation of the principles of installing facilities that provide the best return really should start at the product design stage, so that the product and machinery designer can assess the 'design for manufacture' aspects of each other's designs and produce the most cost-effective solution.

Machinery management has three main objectives, namely:

- To achieve 100 per cent of the quality requirements specified for the product. The machinery must be inherently capable of achieving the desired tolerances and consistency of product quality.
- To achieve the required production capacity and flexibility at the lowest capital and running cost.
- To get the machinery operating reliably at the required production levels quickly and with the minimum of problems.

Machinery designers and manufacturers will be required to respond to more demanding user's specifications, particularly as an increasing number of customers will be implementing TPM programmes in their own companies.

It is important, therefore, that steps are taken to incorporate TPM principles into the specification, design, development, manufacture and installation of machinery.

The diagram in Figure 5.9 illustrates the principles of purchasing and installing TPM 'compatible' machinery throughout the machinery management process from planning through to the hand-over of machinery to production. The main stages of the process are:

1. Evaluation and planning.
2. Specification.
3. Design and operation.

Evaluation and planning

Major projects are generally predictable enough to form part of the business's strategic plan. Therefore, a reasoned analysis of market, product, partnership alternatives and risks will be available and a specific project will be clearly identified as part of the immediate or long-term business objectives. However, it is possible that a project may arise as a result of unforeseen circumstances which is not contained in the current business strategy. Obtain the latest business strategy documents for the business which should include a statement of the policy towards the purchase of new machinery. If these are not available for any reason, a decision must be made as to what the potential project will be assessed against.

Project evaluation

This will normally be in two stages, first, an initial evaluation which assists the business in making the commitment to preparing a more detailed proposal. The initial evaluation should look at the following issues:

1. Relationship of the proposed investment to the strategic plan.
2. Resource requirements against key project milestones.
3. Financial benefit.
4. Risks (technical and financial).
5. Involvement of other partners.

If, after the initial evaluation, the executive of the business decides to proceed, then a Project Manager should be appointed and provided with the necessary resources to lead the work necessary for a more detailed evaluation. Any major factors which could affect the project should be assessed, e.g. rate of exchange, political issues, etc.

If the project would require a step change in the technology level within the company, the business strategy documents should be compared to ensure compatibility. The skills and resources required for the project should be

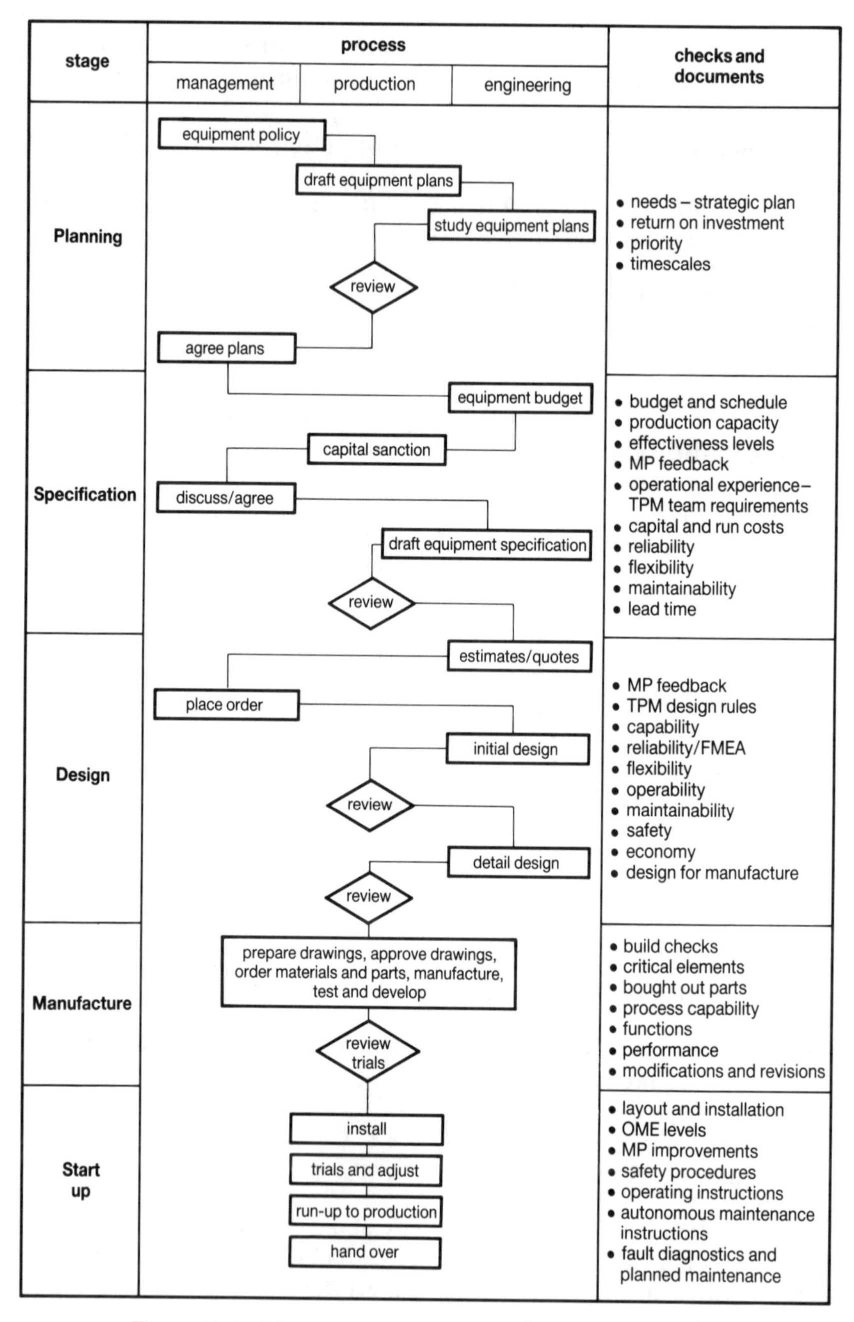

Figure 5.9 The new machinery management process

analysed and compared with those already existing in the company or those proposed by the business strategy documents to ensure that the project would be feasible. Based upon the above considerations, a decision is made as to whether the project will proceed to a more detailed evaluation.

Second, a more detailed evaluation of the alternatives will be carried out. When met with a particular requirement for increased operating capacity which has arisen as a result of:

- Existing machinery being worn out or unable to meet new quality/capacity requirements.
- New manufacturing processes being employed which will enable manufacturing costs to be reduced or product performance to be enhanced.
- New products being introduced.
- Increased demand for existing products.
- New technology being developed which will enable manufacturing costs to be reduced.

there will be many alternatives which need to be evaluated. In design terms, this would be described as 'concept generation and convergence'. Figure 5.10 illustrates the process. Some of the alternatives may include:

Buying out the products, processes or services.

Upgrading existing machinery.

Purchasing new machinery which is:

- Low technology
- High technology
- Fully automated
- Manual
- Somewhere in between

Splitting up the process route and purchasing simpler machinery (i.e. single piece flow or 'nagare' cells).

Through a process of brainstorming, analysis and evaluation the favoured option will be established and further, more detailed analysis, performed on this option.

Before a proposal and a plan can be prepared a marketing plan must be available. Data from this plan can be added to the proposal to enable the team

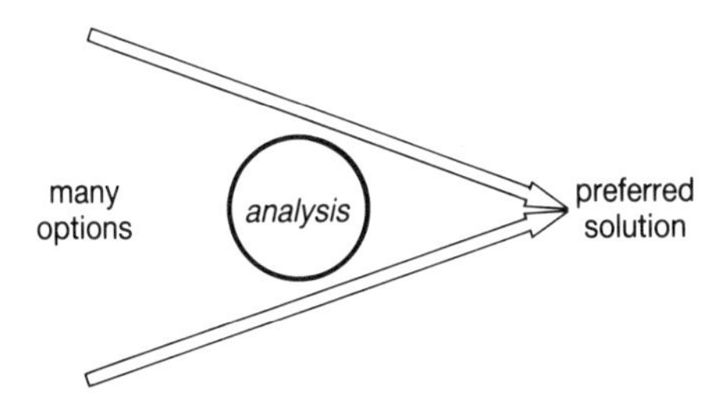

Figure 5.10 Concept generation and convergence

to understand the market (e.g. likely volumes, varieties and probable competitors' reactions) and the customers' requirements and aspirations.

Manufacturing planning

Based on the marketing plan and an input/output analysis for the proposed process model, a capacity planning analysis can be undertaken to look at all aspects of the machinery capacity, labour requirements and capital costs. If this analysis is constructed on a spreadsheet basis a series of 'what if' analyses can be performed to determine the effects of volume and variety changes, automated versus manual systems, changes in machine effectiveness and scrap rates, changes in capital costs and changes in proposed shift patterns.

As well as capacity planning, there is a need to analyse the processes and work handling systems. Initially, all of the processes and options involved should be tabulated and assessed to see which processes are standard and commercially available, which will have to be specifically designed and in which areas new technology is required. In this way, the risks involved can be estimated and steps taken to minimise them. This analysis can then be taken further, again using a spreadsheet, to calculate the numbers and types of work handling containers, the sizes of buffer stocks required to run the system on a 'just in time' basis and to determine the effects of variations in changeover times on the buffer stocks. To complete the manufacturing systems investigation, a process failure modes effects analysis (FMEA) needs to be performed on the system which is used to assess the key areas of risk regarding both failures and defects.

Financial and risk analysis

Calculated overheads, together with data from the capacity planning spreadsheet, can be input into a financial spreadsheet form to yield product costs versus volume, return on capital employed (ROCE), discounted cash flow (DCF) and payback. With the spreadsheet system advocated, sensitivity analyses are easy to undertake by asking a series of 'what if' questions and building macros around the spreadsheets to calculate the effects on product costs, ROCE, DCF and payback automatically. Typical questions will be what if:

volume requirements change?
time slippage occurs?
costs of materials change?
labour rates change?
your competitors react?
machinery performance deteriorates?
overheads change?

Based upon the output of the analyses, a decision on the feasibility of the proposed capital investment can be made.

Risk analysis is a lot more difficult, although risks can be split broadly into two areas – internal and external. Internal risks should come out of the process FMEA and process and work handling analyses, and the consequences assessed. External risks are business risks and should be obvious from the strategic business plans. Further sensitivity analyses can then be performed based on the risk analysis.

Project planning

The planning stage of any project is probably the most critical and requires time and effort to be spent in thinking through all of the actual and potential tasks and problems so that the project can be planned in a structured, logical manner. This also true for a new machinery project which is an important investment for the company. The project plan is a management document which sets out policy, tactics, procedures and objectives. Mistakes and omissions in this document are the principal cause of project failure.

A most important task that has to be undertaken during the planning process is the division of larger projects into smaller sub-projects which are termed work packages. These should be selected on the basis of packages of work which are of a convenient size for the effective monitoring and control of progress and performance against objectives. Each work package should be assigned its own budget and schedule which will relate to the master project budget and schedule. The work package may consist of an element of hardware, software, an ongoing activity or project stage and will depend upon the size, nature and complexity of the project. A schedule of related tasks should be developed for each work package showing major milestones which are, in turn, consistent with the overall project schedule. For more complex work packages, a critical path should be identified. A time-phased resource allocation plan to support the work package should be developed showing manpower and type, services sub-contract support, material, facilities and travel. A spend plan and resource commitment profile should be established in a format which permits detailed tracking.

An overall structure or family tree of work packages is used to maintain the relationship of work packages to each other and to ensure all aspects of the activity are covered.

Application and approval

Provided that the feasibility of the project has been proved, then all of the information generated during the planning and analysis stages should be documented and summarised as part of a proposal document to be submitted to the appropriate senior management of the business and/or to external parties in order to gain approval and financial commitment. The proposal should contain information under these headings:

General

Marketing Information

Engineering Requirements

Manufacturing Requirements

Financial Justification

Project Control

It should aim to justify the proposed capital investment on the basis of its relationship to the business needs, the manufacturing strategy and financial benefit.

Compiling a specification

Once the outline of the project has been proposed, planned and approved in principle, then a specification for the machinery has to be compiled. The specification is a document which clearly and accurately describes the technical requirements for the machinery, including the procedures and criteria by which it will be determined, that the requirements have been met. It will specify not only the required functionality and performance of the machinery but also the standards that must be adhered to, the techniques to be employed to assure the quality of the machinery, operational requirements (input mainly from the TPM teams), constraints and tools, and the features and information to be supplied to assist autonomous maintenance activities and also maintenance personnel.

Elements of the specification

Figure 5.11 illustrates the major elements of a machinery specification which are described explicitly in the document. These can be grouped under the headings of:

1. Functionality – a description of the required functions, processes, throughput, method of operation, finish, etc. (i.e. capability).
2. Reliability – a description of the overall performance of the machinery in terms of its availability to production, overall effectiveness and ease of diagnostics and repair.
3. Environment – the environment within which the machinery has to integrate and perform its particular function.
4. Support – the type and level of support needed to run the machinery up to full production and operate effectively.

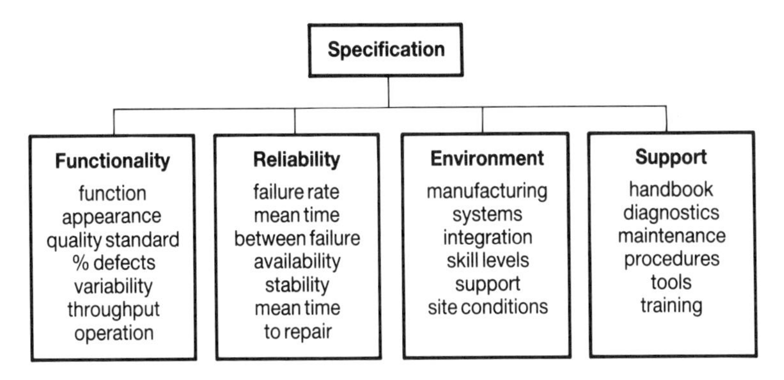

Figure 5.11 Major elements of the specification

It is important that the environmental conditions are stated in the specification as these will affect both the reliability and maintainability of the machinery.

The specification can be used to impose standards, methods and procedures upon the machinery manufacturer which will enhance the reliability and maintainability of the machinery. An example is the insistence that the designer completes an FMEA exercise and discusses this with the user prior to finalising the design. This does not automatically build reliability into the design, but it acts as a focus for reliability assessment during the design process. Similarly, the imposition of maintainability standards and checklist procedures will ensure that the subject has been given due consideration.

In essence, the machinery specification can be utilised to specify not only function and performance, but also the means by which aspects of the design and manufacture of the machinery will be carried out.

Contractual support

The Terms and Conditions of contract for purchase of machinery may be regarded as the tools used to ensure that the manufacturer provides machinery which meets the specification, timescales and price. Similarly, from the manufacturer's point of view they clearly define the extent of his supply and contractual obligations. Stage payments can be tied to the provision of documentary and physical evidence that the specification has been achieved and that measures have been complied with.

The required level of overall effectiveness (OME) for the machinery can be specified but the means of measuring this and whether this applies to the overall machinery and/or its elements or component level has to be specified. Also the mean time to repair (MTTR) for the machinery can be specified but, similarly, the means of measuring and recording this figure have to be determined. This can be incorporated into the contract in the form of a warranty which states that a certain percentage overall effectiveness figure has to be attained for a specified period of time in order for final acceptance of the machinery to be achieved.

A two-stage acceptance can be very useful in separating acceptance of the capability of machinery from acceptance of levels of overall effectiveness. In the latter case, a modified form of up-time warranty is a very useful method of tying final acceptance (and payment) to the achievement of an agreed percentage availability over a reasonable period of time. Many purchasers use this method which can be very effective, provided that the means of measuring and reporting breakdowns are clearly stated and implemented.

The provision of spare parts, test equipment, manpower, transport and the maintenance of both spares and test equipment is a responsibility which may be divided between supplier and purchaser or may be assigned to either. These responsibilities must be described in the contract along with training requirements and the levels of skills needed to support the machinery.

Acceptance criteria

The machinery specification and contractual conditions must include a statement of the acceptance criteria pertaining to the application. Many problems arise in machinery purchases as a direct result of the lack of, or badly constructed, acceptance criteria. A clear and unambiguous statement of the standards and tests that have to be achieved by the machinery is paramount. It is preferable to define various acceptance stages and the standards/tests that have to be performed and achieved at each stage. Stage payments may be linked to the acceptance stages.

Well thought out, reasonable and unambiguous acceptance criteria enable the purchaser to be clear about what he is gaining at each stage and allow the supplier to understand his responsibilities at each stage. It is a grave mistake for either party to conclude a contract without the acceptance criteria being agreed as part of that contract. Do not be tempted to state 'Acceptance Criteria To Be Agreed' – what if you cannot agree?

To understand what needs to be included in the acceptance criteria it is useful to define the quality of machinery and state how this can be measured. Figure 5.12 illustrates the factors which measure and determine the quality of machinery. Conformity and reliability are measured by the ability of the machinery to do the following:

- Perform the specified functions.
- Perform the specified processes.
- Achieve the specified throughput rates.
- Achieve a consistent output.
- Display acceptable levels of variability and % scrap levels.
- Achieve the specified level of OME.

This may be referred to as machinery 'capability'. What is normally referred to as machinery 'reliability' can be assessed if the machinery operates as follows:

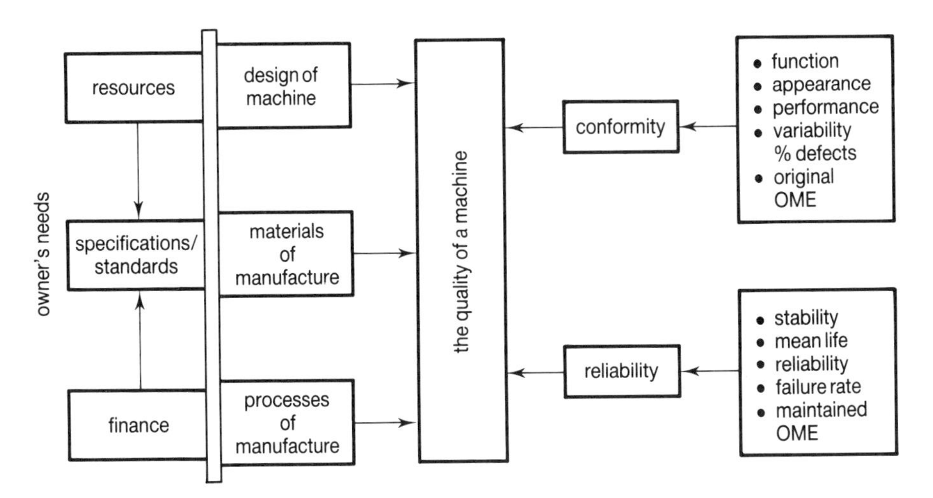

Figure 5.12 Machinery quality

- Provides reliability of output.
- Operates in a stable manner.
- Achieves a low rate of failure.
- Consistently achieves the specified level of OME.
- Achieves the specified mean time to repair.

The overall quality of machinery is measured against the user's needs and expectations, most of which will be documented in the specification. This is where the TPM team in whose area the machinery will eventually operate can provide valuable input to the specification. There will, however, be implied needs/wants and more subliminal expectations which need to be addressed if the user is to perceive value for money. The functions, processes, quality and rate of output, resource requirements, finance available, etc. will be the major user inputs to the machinery designer.

Supplier relationships

Once the supplier has been selected and an order placed for the machinery, it is then necessary to make arrangements with the supplier to ensure that delivery dates, costs and the specification are met.

The supplier must, having already submitted an indication of project timing, now produce a critical path analysis giving start and finish dates, with each aspect of the work timed and the critical path items highlighted. This will ensure that all the supplier's departments have been consulted to establish a timescale for their part of the project. An initial meeting should be set up to go through the specification and terms of contract with the supplier to ensure nothing is overlooked. Also samples of the product with drawings must be supplied to the supplier as an aid to machinery design and testing. It is also important that a good working relationship is established between supplier and purchaser to aid technical discussions.

Maintaining the project team structure is essential, and changing the staff who are dealing with a particular supplier can destroy this relationship, with the resulting loss of supplier confidence and the possibility of confusion arising from the fact that the new engineer is not totally *au fait* with the project or supplier.

Regular visits to the supplier are absolutely essential, along with an updated milestone or barchart submitted by the supplier. This will ensure he/she is keeping to time and that the specification is being adhered to. The effects of a poor specification are particularly evident at this stage, as items incorrectly specified will increase costs in the form of extras. It may be that the supplier submitted an unusually low quote to gain the order and will seek to recover costs through overcharging for extras.

The supplier may well also be sub-contracting and these companies will require monitoring. It is especially important that a written commitment to timing is received from them. Suppose the suppliers had only a verbal delivery date quoted to them from their sub-contractors and, upon activating the delivery, find that due to extra workload, the sub-contractors cannot deliver on time. The result is that the whole project may be affected if the activity is on the critical path. The initial supplier should provide a list of major bought-out-items. The stage payment system should then be utilised by the supplier to order these bought-out-items as soon as possible, not as late as possible.

If written in the contract, the machinery should be built and tested at the supplier's premises, prior to delivery to site. This can reduce costs and time by minimising supplier's engineer's visits to site, probably to fix a minor item.

Design and operation

It is becoming apparent that as more and more operating companies progress in the implementation of TPM, they will increasingly use overall effectiveness as their main measure of performance for machinery. Future user specifications will include OME figures as part of the acceptance criteria, and machinery purchasers will also require evidence that the design has been evaluated with respect to OME and measures put in place to reduce the six big losses. Machinery will be expected to achieve the specified level of OME and to maintain that level of performance throughout its life.

TPM design principles

Throughout the machinery design process there is a need to ensure that the necessary steps are taken to achieve the required levels of overall effectiveness, and also to provide machinery which is 'TPM friendly' and incorporates design features which do the following:

- Make the machinery inherently reliable.
- Enable its reliability to be maintained easily.
- Make the machinery inherently capable.
- Enable its capability to be maintained easily.
- Enable the machinery to be operated without danger or difficulty.
- Enable changeovers from product to product to be quick and easy with the minimum of setting.
- Make cleaning quick and easy.
- Give clear, visual indications of the state of the machinery.
- Enable lubrication procedures to be carried out quickly and easily.
- Give easy access for checks to be carried out on critical elements of the machinery.
- Offer improvements as a result of information gained concerning the problems with existing machinery and the operational improvements made to overcome these problems.

This last feature can be achieved only if mechanisms are put in place to feed back the experiences of machinery users and specifically TPM teams and then to ensure that these experiences have been considered and, where possible, incorporated into the design of subsequent machinery. This is at the very heart of what is often referred to as 'Maintenance Prevention' (MP) design, and relies upon a spirit of co-operation between the machinery user and manufacturer and the existence of simple and practical systems for communication and analysis. The diagram in Figure 5.13 shows the main elements of a maintenance prevention design system.

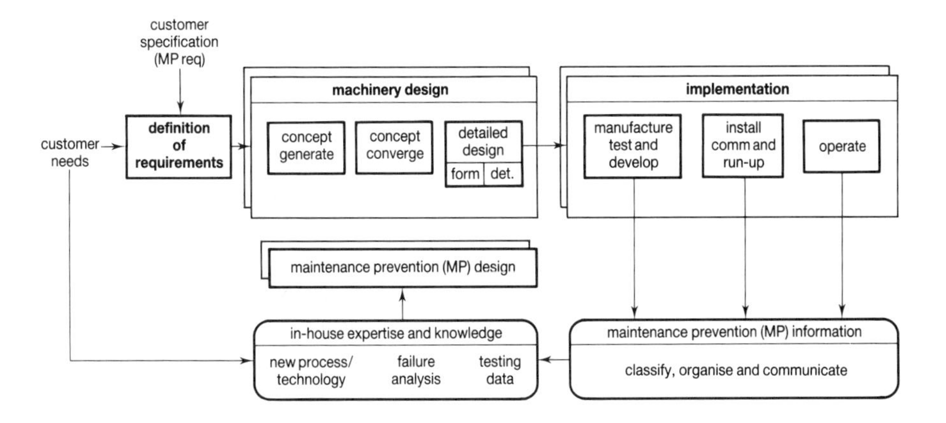

Figure 5.13 A maintenance prevention design system

Maintenance Prevention design

A Maintenance Prevention (MP) design system is created in stages. During the implementation stages of machinery design, i.e. when the machinery is manufactured, tested and developed, installed, commissioned and 'run up' to full production levels and operated thereafter, maintenance prevention information is generated by both the machinery user and manufacturer. This information has to be gathered, organised into categories, machinery type, defect type, etc. and communicated to the machinery manufacturer.

The MP information thus communicated has to be analysed and added to the machinery manufacturer's in-house information concerning the following:

- New technology developments.
- New process developments.
- Failure analysis of other information gathered.
- Data from testing of machinery and machine elements.

The combined information has to be introduced into the design process by means of checklists, guidelines, training material, standards and design reviews. This will then be used throughout the machinery design stages to ensure that the resulting design incorporates as many good features as possible and achieves a high level of overall effectiveness.

MP design information may also be received from customer surveys and customer specifications. This is particularly true where TPM teams have been established on customer sites. In the latter case, where additional TPM design features are to be incorporated into the machinery design, it may be necessary for the designer to discuss these with the customer's TPM team as part of the initial design review.

There are typically five main areas where problems can arise in the MP design system, namely:

1. Maintenance reporting. The TPM team/maintenance personnel will be required to keep records concerning breakdowns, repairs, preventive maintenance or other activity. These records will need to be analysed by machinery designers. Problems arise when the information is inadequate or unclear.
2. MP information collection. The information passed to the machinery designer may not be clear, the order of priority and severity of the problem and its effects not shown and the information mixed up.
3. Design guidelines and standards. Quite often, the MP information is put to use only once, for specific machinery, and not applied to other designs where it would be relevant. Or, alternatively, general design guidelines or standards are too vague and open to interpretation.
4. Checklists. Can be very useful, but are not foolproof and can become out of date or irrelevant over time.
5. Interpretation of MP design. Vague methods of assessing the machinery

design and how well it incorporates the MP features lead to different interpretations and standards.

In order to overcome the problems detailed above, the following mechanisms should be put in place:

1. Use a standard 'MP feedback' form which gathers all of the relevant information and can be easily sorted into categories. TPM teams/ maintenance personnel will need to be trained in how to gather the relevant information and fill in the form.
2. A simple administration system for categorising and filing MP information must be established. A computer data base program can be used to provide easy access and sorting of information.
3. Produce an MP design guide book for designers and provide means of continually updating the guide in the light of more MP information. More specific design guides can be produced for specific machines or machine types.
4. Produce checklists for MP designers to use at each stage in the design process. These may be incorporated into the MP design guide book. Checklists for each aspect of machinery design and to cover all disciplines can be compiled and means of continually updating and improving them must be provided.

Designing effective machinery

In addition and complementary to the use of MP information in machinery design, there are a number of analysis techniques and design approaches that can help to ensure that designs are inherently effective. The requirement is to design machinery that incorporates the following features:

- Reliability.
- Consistent operating performance.
- Maintainability.
- Consistent process capability.
- Flexibility.
- Safety.

Certain approaches and techniques can be used by the machinery designer to ensure the effectiveness of the design within the constraints placed upon him or her. Failure Modes Effects Analysis (FMEA) has been used for product and process analysis for many years. It is essentially a risk analysis of the design which is meant to show up any high risk areas, so that action can be taken to reduce the risk either through design changes or operational procedures.

The form provided in Appendix 2 can be used for the analysis of machinery at the design stage. It asks:

- What are the major components of the machine?

- What is their function?
- How could they fail?
- Why could they fail?
- What planned maintenance would need to be carried out?
- How long for and how often (estimated)?
- What is the risk level of failure occurring?
- Corrective action planned?
- Revised risk level as a result of changes?

Problem-solving/fault analysis techniques such as those described in section 5.6 can be applied to the design and development of machinery.

Throughout the machinery design process, the emphasis is placed upon the achievement of an inherently reliable and capable design which achieves its functional and performance requirements. In order to maintain high levels of machinery effectiveness, it is necessary to ensure operational reliability, i.e. a design that provides features which assist the operator/setter to ensure that failures and defects do not occur. Lack of consideration of operational reliability can lead to the following problems:

- Failure to clarify correct operating conditions.
- Assumptions that experienced operators will set the machinery correctly.
- A range of operating conditions that are either too narrow or too wide.
- Operating procedures that are unclear and difficult to carry out.
- Changes in operating conditions are not detected until defects or breakdowns occur.
- Conditions are somewhat unstable and fine tuning is often required.
- Changes in operating conditions due to deterioration are hard to detect and difficult to restore.

Understanding all of the process parameters which affect the quality of the finished product is an important prerequisite to be able to quantify and define the correct operating conditions. The way in which machinery components affect each process parameter, how sensitive this is and how critical it is to product quality can be analysed using the FMEA technique, but concentrated on quality issues. The results of such an analysis will form the basis of the operating procedures and also will direct engineering design effort towards making the more critical components more robust and stable in operation, easier to set up, easier to check and/or incorporate automatic detection of changes and easier to restore and/or replace.

■ 5.9 Winning the support of people within the company

Initially, many people within the company will be highly sceptical about TPM and will question whether it can succeed and provide benefits to themselves or the business. They will question the validity of any claims about the benefits of TPM, as though some large-scale deception has been employed in all of the companies that have implemented TPM. My experience is that the factory floor personnel, once they have understood what TPM is all about, really want it to happen, but have little faith in their managers and, because of that, doubt very much whether TPM can succeed. Managers, on the other hand, take much more convincing about the efficacy of TPM and, although many of them may be suffering from the 'typical scenario' as described in Chapter 2, they are apprehensive about what changes may mean to them.

Simply talking about TPM will not convince people or break down any of the barriers to its successful implementation. That is why it is essential to follow the introductory training and communications sessions very quickly with the pilot implementation of autonomous maintenance. Once people within the company have seen what TPM is in practice, have experienced its benefits and seen or taken part in some of the activities, they begin to understand the implications for the company and for themselves. Something has to be seen to happen quickly so that TPM can begin to gain credibility and to extinguish any lingering doubts and scepticism.

Rather than attempt to introduce all of the components of TPM as described in Chapter 3 and illustrated in Figure 3.1, it is advisable to start with relatively simple TPM activities related to autonomous maintenance and creating a clean and tidy workplace (CAN DO disciplines). The reader should not expect to convert everyone in the company to wholehearted TPM supporters 'overnight'. This is about changing people's attitudes, values, custom and practice and will take time to achieve. The length of time will vary greatly from individual to individual. Some people will very quickly become supporters of TPM, some will take years, a few may never change as they have become so set in their ways. It will take years rather than months to get the majority of people 'on board' and irreversibly to change the culture of the company to one which truly embraces the philosophy of TPM.

■ 5.10 Applying TPM in different industries and areas

The principles and overall philosophy of TPM can be applied to operating companies in all major industry sectors that undertake the manufacture of many and varied products. They can also be applied to many areas within those companies as discussed in Chapter 3. The main difference between implementing TPM on the factory floor in an automotive components manufacturing company or in the sales office area of a manufacturer of sophisticated electronic equipment is, therefore, not the principles and philosophy of TPM, but the way in which it is applied. The emphasis which is placed upon certain components of TPM, the method of introducing it, the pace of implementation, etc. will vary from situation to situation. Over many years, I have been closely involved in training personnel and supporting the introduction of TPM in many different operating companies and areas of those companies, and each time, the implementation has needed to be subtly different. Even in operating companies which have two different sites which manufacture the same product using the same machinery and processes, the approach to TPM has been different. The need for a different approach is due to the fact that every company and operating site has the following features:

- Its own background which has led to the establishment of its own unique culture.
- Personnel who are native to that part of the world, country and area and have their own local culture.
- Custom and practice, terminology and methods which are unique to their industry sector.
- Its own unique set of people, skills and experience.
- Its own particular environment, site layout and constraints.
- Different machinery, equipment and general facilities.
- Organisational structure and reporting relationships.
- Systems and procedures, etc.

Figure 5.14 shows the typical characteristics encountered in different production environments. This is meant to be purely indicative of the situation which often prevails in each of the four operating company types and also how difficult it is to initially introduce TPM. There are many exceptions, but the table provides some general guidance based upon many years of experience.

How does one implement TPM in these different environments and also in different areas of the company? The approach, points of emphasis and

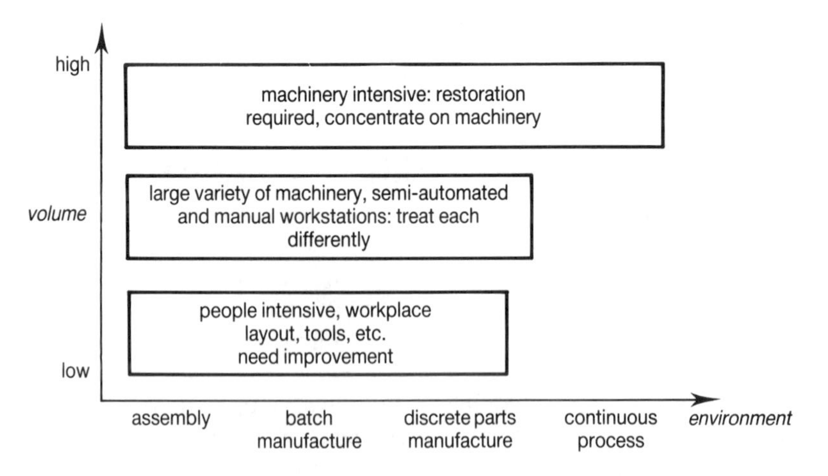

Figure 5.14 Typical characteristics of different production environments

guides to implementation in a number of environments and areas are discussed below:

Medium- to high-volume manufacturing

This type of company (for example, the automotive industry) has a great deal of machinery, most of which is in poor condition. The introduction of autonomous maintenance must commence with the identification of faults and operating problems of which there will usually be many. A great deal of effort will, therefore, be needed to restore machinery, and because operator skills will be variable, but generally low, and because of custom and practice, there will be a reluctance by members of the TPM team to work on machinery and fix faults. A great deal of support will be required from the maintenance department early in the implementation as a result of this, but team members should be encouraged to assist maintenance personnel when they are carrying out the work. This will gradually achieve a better production/maintenance relationship and understanding and will also start to give operators a greater awareness of how the machinery actually works.

This kind of environment usually has both a strong union representation and a traditional 'them and us' relationship between factory floor personnel and their managers. Communication is the key in this situation, both prior to launching TPM and during implementation, and people will be very sensitive initially.

The use of TPM co-ordinators is also crucial in this environment as the prevailing 'production at all costs' attitude will quickly prevent TPM activities and working sessions from proceeding, unless the team is supported properly. Very strong high level commitment is necessary as production pressures will be very great and a reasonable amount of money will have to be found to finance the TPM programme.

Overall effectiveness levels will vary, but will normally be quite low, around 50 per cent to 60 per cent, this should be improved quite quickly. The general state of machinery will be quite poor, a great deal of dirt, oil and other process debris is to be expected and people have become so used to this that they have to be convinced that the workplace can be made clean and tidy. As the loading of machinery is quite high, two or three shifts will often be employed and communication across shifts is very important. Run TPM teams on different shifts, and particularly make sure that the night shift is not left out.

Continuous process manufacturing

The food industry or the steel industry are examples of this kind of manufacturing. The critical machinery or vessels in a process environment out of necessity tend to be kept in a reasonable condition, and also there are areas that the process or regulations dictate are kept very clean. Unfortunately, other machinery is often neglected and the rest of the workplace can be dirty and untidy. Autonomous maintenance should be introduced, but it will be limited initially by the skills and confidence of operators. This is mainly because the divide between engineering and operations has traditionally been quite substantial in such an environment with engineering personnel being viewed as somewhat 'elite' and having responsibility for all aspects of machinery on the factory floor. Experience shows, however, that operational personnel often have higher skill levels than anticipated. It is necessary to work at the operations/maintenance relationship by doing the following:

1. Including an engineer in the TPM team.
2. Setting up focused improvement projects which are run by engineering, but have operational personnel on the team.

There needs to be a senior management commitment to the principle of engineering and operational personnel working closely together and gradually moving towards the roles which were described in Chapter 3, section 3.2 and illustrated in Figure 3.3. Some basic skills training will be required for operations personnel as the TPM implementation progresses and engineering personnel can provide a valuable input to the training.

Overall effectiveness levels are often quite high, 80 per cent to 90 per cent is not unusual, but because the machinery is normally very highly loaded (24 hours/day, 7 days/week is often the case) and the value added by the process is quite substantial, an increase of only a few percentage points releases a great deal of additional earning capacity. Scheduling TPM activities and working sessions can be particularly difficult in the process environment, but as regular 'cleaning time' is often scheduled it can be convenient to carry out TPM either during this period or by extending it.

TPM pilots should be run in both the main processing areas of the company and in the packaging or collating areas as they are very different. Communications, particularly during the early stages of implementation need to be kept very simple and basic as many personnel do not have any engineering knowledge. It will also help if the TPM co-ordinators chosen have some engineering knowledge already or are quickly trained.

Because people often work different shifts, cross-shift communications are very important, and the way in which a team is initially set up and evolves will be very much affected by the shift pattern.

Low-volume batch manufacturing

Machine shops and tool rooms are examples of this kind of manufacturing. In this environment, it is usual to encounter many different machinery types, sizes and ages and it can be difficult to choose a suitable pilot. I usually look for a group of similar machinery and start by allowing the team to choose one or two which are not the best or the worst but somewhere in between. Machinery is usually in poor condition, but there will sometimes be individuals who do look after 'their' machinery. When scrutinised, the workplace will usually be dirty, oily and untidy, even if it looks reasonable from the outside. Skill levels of factory floor personnel will tend to be quite high, and with a little encouragement the TPM team members will be able to rectify many machinery faults and carry out a fairly high level of autonomous maintenance. It is usual to find that the number of maintenance personnel is small. Some companies do not have any and rely upon their operations people, supplemented by outside contractors as required. Maintenance personnel will normally be very supportive of TPM and will regard it in a very positive way.

Overall effectiveness levels will vary from 40 per cent to 80 per cent and can usually be increased to 90 per cent plus, through autonomous maintenance activities. Because of the low volumes encountered in this environment, it will be necessary to calculate percentage performance and percentage quality, based upon the hours lost due to stoppages and reduced speed for the former and parts being scrapped or re-worked for the latter.

Loading varies with some machinery being very highly loaded and others being used only occasionally. The difference between effectiveness and utilisation has to be stressed to ensure that all machinery is eventually included in the TPM programme. Sometimes, even lightly loaded machinery can cost a great deal in spares and service visits. It is advisable to run pilots on different types of machinery and also to spend time on jigs, fixtures, tooling storage and organisation as these can be important issues in this environment.

Low-volume assembly

Invariably this environment (for example, the assembly of machinery) is orientated around the manual assembly of parts and some products require very specialised fitting skills. The facilities used consist of small items of equipment, lifting equipment, fixtures, tools and racks. It is important to identify all of the facilities required to complete an assembly, right down to hand tools, and then to investigate whether they are generally available and what state they are in. Personnel are usually highly skilled, and the restoration of facilities and the requirements for autonomous maintenance are well within their abilities. They will be used to carrying out a great deal of repair work themselves and, if there are any, the number of maintenance personnel will usually be small. The implementation of 'CAN DO' disciplines should be emphasised in this environment as the majority of losses are incurred because people are looking for something. A great deal of time should be spent on sorting out fixtures, tooling, materials, etc. and arranging racks, cupboards, shadow boards and 'parking places' in the assembly area. The shortage of parts is usually an issue and may best be handled by setting up a focused improvement project to concentrate on the supply of parts. Although it is necessary to undertake an investigation of any faults and problems, the number of these should not be great and the most significant exercise will be the generation of area improvement ideas.

Overall effectiveness can be measured by using a modified version for overall assembly effectiveness which measures hours lost per assembly. This calculates percentage availability, performance and quality based upon the assembly hours lost due to these types of delays. The overall measure of performance can be based upon the average number of hours taken to complete an assembly and TPM is aimed at reducing these. The TPM team may start fairly small, but will gradually expand as more small teams which are set up to concentrate on certain aspects of assembly are added. The actions that the initial team decide upon which affect shared facilities need to be communicated to all users and, eventually, a rota system for autonomous maintenance activities can be organised.

The attitude to quality is sometimes quite poor and there is a need to emphasise the importance of quality systems and equipment. This equipment should be marked clearly and stored in designated areas.

Medium- to high-volume assembly

A combination of automatic assembly machinery, semi-automatic machinery and hand assembly is often encountered in this environment (such as the assembly of office equipment, for example). Generally, the machinery is not well looked after and many minor stoppages happen, to the extent that automatic machinery often has a minder standing next to it. Quality is often variable and is an issue of great concern, and usually poor parts supply is blamed. It is important to concentrate on all of the facilities used from the automatic machinery to the hand tools used for manual assembly. Operators are often low skilled and their understanding of engineering principles is assumed to be very low; however, this is not always the case.

Overall effectiveness levels do vary from 40 per cent to 85 per cent, and it is fairly straightforward to use the standard overall effectiveness calculation in this environment. Collecting overall effectiveness information will often expose the need for focused improvement projects, primarily associated with quality and/or performance problems.

TPM pilots should be run on both automatic machinery and hand assembly stations to show the differences, and the latter will very much concentrate on operator problems (ergonomic issues).

Initial communications are very important as rumours seem to spread quickly in this environment and local TPM co-ordinators are necessary to keep the momentum going and support the team. There are often few maintenance personnel with many maintenance activities being carried out by craftsmen who also change over the machinery and work stations. Some training in the engineering basics will probably need to be considered when implementing autonomous maintenance.

TPM in the office

The starting point is to investigate the process which is carried out within the particular office area and to list each activity which the team members undertake. Against each activity, the facilities which are used should be listed. Any faults or operating problems can be highlighted and tagged as in a production environment, but there will probably not be too many of these. Most activity will centre around 'CAN DO' disciplines and area improvements.

Because most of the equipment used consists of electronics (such as telephones, computers, facsimiles, etc.) there is a limited amount of operator-based maintenance that can be carried out. The service arrangements for the equipment can, however, be investigated and improved if any problems exist. Surprisingly, cleanliness is often an issue in an office environment and equipment and the general office area can get very dusty. On some occasions the office cleaning arrangements are questioned and the team members can end up cleaning the area themselves. The team should be encouraged to throw away old files and any other unused items and to establish a place for everything and everything in its place.

The main measure of performance will be based upon the average time taken to achieve the defined process and thus, TPM activities will be aimed at reducing the average lead time, eliminating errors and improving the working environment.

TPM in Research and Development

A wide range of facilities are usually employed in a Research and Development environment, mainly for analysis, testing and development. It is best to treat them in the same way as production facilities, even though their loading time will usually be quite small. The emphasis should be that the need for effective facilities is still relevant in this environment as development programmes can be severely disrupted if facilities are not performing correctly. Overall effectiveness can be measured, although it can be difficult to monitor performance due to the 'stop/start' nature of development work.

If operating companies are embracing the TPM philosophy, it is important that their research and development support areas are also applying TPM. This helps working relationships, aids understanding and also emphasises the validity of development programmes. 'CAN DO' disciplines and autonomous maintenance activities can be applied successfully in laboratory and workshop areas.

6 Consolidating and Developing TPM

During the time that TPM pilot areas are being implemented, autonomous maintenance is being introduced and planned maintenance is being developed, there is a great deal of activity and TPM has a high profile. When the initial restoration and development of autonomous maintenance procedures have been completed, the 'spotlight' may then move from a particular area of the company to somewhere else. It is at this point when TPM may run out of steam. The enthusiasm of the team, and most importantly, local line management, can decline and working practices revert to the 'bad old ways'. It is essential that mechanisms are put in place which will not allow this decline to happen and will ensure that TPM activities will continue and progress.

6.1 Measuring improvements

Although the TPM activities described in the previous chapters will have a beneficial effect upon the pilot area and also on the personnel involved, it is difficult to quantify the benefit to the business without some visible and tangible measures of performance. For this reason it is necessary to install overall effectiveness and to measure, plot and display the figures for at least the most important machinery, and at best, all of the machinery in the area.

The improvement in overall effectiveness which normally accompanies the implementation of TPM can be directly related to the income that is earned by machinery and it should be possible to calculate actual savings as described in Chapter 4, section 4.6. Other measures of performance such as the number of faults rectified, number of suggestions made, etc. can be useful indicators to which area personnel can easily relate, but they can only be translated into savings via the overall effectiveness calculation.

Highly visible measures of performance are, therefore, very necessary indicators that TPM activities are being carried out in an area and should be linked to the personal assessment of line management in that area. The achievement of the objectives for all personnel working within the area should be measured by the following:

- Improvement of overall effectiveness.
- Number of improvements made.
- Amount of time spent on TPM activities.
- Amount of training and new skills obtained.
- Improvement of the workplace environment.

Once autonomous maintenance activities are under way, it is necessary to display the relevant measures of performance on a notice board in the area. Figure 5.3 shows a typical overall effectiveness chart which should be completed and displayed. An 'effectogram' as described in Figure 3.5 can similarly be completed and displayed. The number of faults rectified and improvements made can be displayed in histogram format and 'before and after' photographs can be very effective. Most Japanese factories have many square metres of notice boards around the factory displaying all kinds of information, a great deal of which is kept up to date by the TPM teams and local supervisors. The key to displaying such information is that it must be regularly updated or it will very soon lose its credibility.

6.2 Developing a TPM programme

In order to ensure that autonomous maintenance and other TPM activities are carried out, there are certain mechanisms which have to be put in place within the business. These have to become part of the everyday working practices of the area and will include:

- Regular TPM team meetings with a set agenda.
- Ongoing collection of overall effectiveness information.
- Weekly calculation and plotting of overall effectiveness.
- Ticklist for autonomous maintenance procedures.
- Fault/problem reporting and recording system (including tagging).

Figure 6.1 shows the mechanisms that are required to support TPM in the company and how they revolve around the regular TPM team meeting.

A vitally important element of support which has to be put into place if a TPM implementation is to succeed is the appointment of TPM co-ordinators or team leaders. As TPM spreads within the company, a network of co-ordinators needs to be established from individual team co-ordinators up to an overall co-ordinator for the company. The individuals concerned need to be inspired by the TPM philosophy and committed to its implementation. The role of the TPM team co-ordinator includes the following tasks:

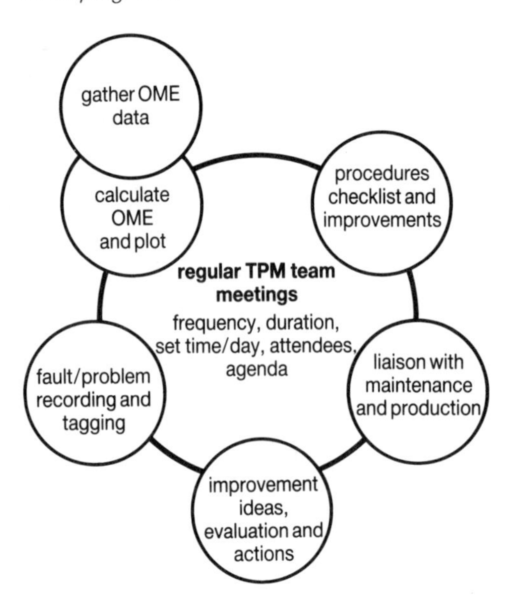

Figure 6.1 TPM support mechanisms

- Helping and organising team members.
- Co-ordinating TPM activities and working sessions with production.
- Co-ordinating TPM activities with maintenance.
- Organising outside help, special resources, parts and materials.
- Arranging and facilitating team meetings.
- Liaising with other TPM team co-ordinators.
- Reporting progress to the area or company TPM co-ordinator.

It will be necessary to train the co-ordinators both in the principles, philosophy and practices of TPM in detail and in how to lead and co-ordinate a TPM team. Usually, the team co-ordinator will emerge as the natural leader of the team, but the role is not necessarily assigned to the local supervisor. In some situations it is appropriate for the team to elect their own co-ordinator and this is to be encouraged. The company TPM co-ordinator needs to be at senior manager level, reporting to the managing director or equivalent so that he/she has a sufficient level of authority to drive the TPM programme through if necessary. Part of the TPM implementation process is to review constantly where the team and the business are and to keep increasing the standards. For example, the initial level of cleanliness that the team achieves in a pilot area will be a considerable improvement on pre-TPM levels, but as the implementation progresses it will be regarded as 'not good enough' and standards lifted progressively. The constant need to improve has to be ingrained into the whole culture of the business.

Once the pilot has been completed, it is an appropriate time to decide whether TPM should be implemented across the whole business. Provided that the pilot has been run properly and that the 'philosophical' changes have begun, then the answer will almost certainly be yes. At this point it will be necessary to develop a TPM implementation programme for the business which will probably span two to three years. The overall objective of the programme will be to implement all of the components of TPM, as detailed in Chapter 3 and illustrated in Figure 3.1, along with changing the culture of the business as discussed in Chapter 3, section 3.1. The TPM programme should include the following features:

- Spreading 'simple' TPM to all other cells/areas.
- Institutionalising the 'simple' TPM activities in all areas.
- Introducing other TPM components.
- Training and developing factory floor personnel.
- Training and developing maintenance personnel.
- Considering, introducing/changing incentive schemes.
- Installing overall effectiveness as a main business measure of performance.
- Measuring the benefits arising from TPM.
- On-going communications.

Personnel at all levels of the business should be consulted when the programme is being compiled and the agreed programme communicated throughout the business.

The reader should note that it is very difficult to compile an overall TPM programme without first implementing one or more TPM pilots for the following reasons:

- The approach to TPM implementation will vary from company to company and area to area, and the best approach cannot be developed until TPM has actually been applied in practice.
- The amount of time, resources and finance required to implement autonomous maintenance can be established only as a result of undertaking a pilot or pilots. This allows a realistic programme to be compiled within the constraints of company resources.
- Through the experience of running a pilot or pilots, the pace of change which can reasonably be expected becomes apparent.
- Only through the experience of running a pilot do the implications for managers and the company as a whole become clear.

I would advise against spending a great deal of time and money undertaking studies and compiling detailed TPM programmes prior to introducing TPM to the company.

After completion of a successful pilot, the TPM programme should become an agenda item at the appropriate senior management meetings. The company TPM co-ordinator should be required to report regularly on progress and the programme should be integrated into the overall business plan.

■ 6.3 Training and people development

Although initially the business may need outside parties to provide presentations, training and material and to help set up and run the pilot(s), compile the programme, etc. many businesses will wish to become 'self-sufficient' after the initial introduction of TPM. This will require someone on the site to take on the role of the 'TPM champion' who will drive and co-ordinate the TPM teams and activities across the business. The TPM champion can be used to develop specific material and to provide ongoing training and support and larger businesses may need to 'grow' a network of TPM trainers/facilitators.

One of the underlying aims of the TPM implementation programme should be to develop shop floor personnel and allow them to realise their full potential. Training will need to be provided for new skills, an understanding of how machinery works and how to carry out specific TPM tasks. The latter can be supported and/or provided by maintenance personnel. The aim should be to provide 'on the job' training where possible as this has proved to be most effective in the context of TPM.

Operator and craftsman training

As part of the TPM team preparation and launch, all members of the team will need to be provided with training in TPM principles and practices, most of which can be reinforced with 'on the job' examples. TPM training for factory floor personnel is much more effective when related to their particular working area and shown in practical terms. The checklist in Figure 6.2 shows a typical list of subjects that need to be included in individual training and development plans. Where personnel have already completed an apprenticeship or approved training course, then the appropriate block can be shaded in and where no training has been received then this can be organised. It is normal for machinery manufacturers to provide training for operators, craftsmen and maintenance craftsmen as part of the package for new machinery. Where personnel are being assigned to machinery that they are not familiar with, then on-the-job training needs to be provided by those who are experienced in the operation, setting up and maintenance of the machinery.

Maintenance personnel training

The training of maintenance personnel should include the principles and practices of TPM and also training in the different types of maintenance techniques available which support preventive and predictive maintenance. Figure 6.3 provides a list of the subjects which should be included in the training plans for maintenance personnel.

Subject	Operators									Craftsmen								
Safety																		
Quality basics																		
Handling																		
Basic mechanical principles																		
Engineering materials																		
Fasteners																		
Using hand tools																		
Using inspection equipment																		
Basic drive components																		
Basic pneumatics																		
Basic hydraulics																		
Basic electrics																		
M/C services																		
Lubrication																		
Machinery/equipment operation, set and adjustment																		

Figure 6.2 Training plan for operators and craftsmen

■ 6.4 The business implications of TPM

It has been explained that TPM is very much a team-driven initiative which relies upon the motivation and participation of all operational personnel. There are, however, business implications attached to the implementation of TPM which need to be understood.

First, TPM is aimed at effective operations not just efficient operations where the goal has traditionally been to produce as much as possible, in the shortest possible time at the lowest cost, irrespective of whether the goods can be sold, stored or meet quality and market requirements. The business has to come to terms with the characteristics of effective manufacturing and realise

Subject	Maintenance									Craftsmen								
Safety																		
Quality maintenance																		
Handling																		
Mechanical principles																		
Practical electrics																		
Electro/pneumatics																		
Hydraulics																		
Machine controls																		
PLC programming and operation																		
CNC controls																		
Industrial electronics																		
Pipefitting																		
Welding																		
Instrumentation																		
Pumps and compressors																		

Figure 6.3 Training plan for maintenance personnel

that machinery does lie idle at times, operators can be involved in activities other than operating machinery and maintenance activities are essential.

Second, at the very heart of TPM lies the encouragement and enhancement of factory floor personnel. The traditional relationships between factory floor and management have to change from distrust and confrontation to mutual understanding, trust and team working. This will not happen overnight, but through time will happen for TPM and as a result of TPM. The implications for managers within the company is that their role and attitudes will have to change and their responsibilities become as follows:

- Leading and supporting people.
- Helping and encouraging the TPM teams.

- Allowing and encouraging TPM teams to become more self-sufficient, to take decisions and organise their own activities.
- Developing the potential of their people.
- Looking for long-term improvements rather than short-term fixes.
- Communicating with people, i.e. listening as well as talking, looking for good ideas and having respect for their team.

If managers can begin to change their role then their credibility with personnel and enjoyment of the job will increase dramatically.

TPM is orientated around facilities and the people who operate, set up and maintain those facilities and it will have a profound effect upon the whole business.

7 Conclusion

Total Productive Maintenance is poised to bring about a revolution in the way in which companies operate. It will not be a rapid, high profile revolution, but instead a gradual and irreversible 'creeping' change. The TPM journey begins with a spark of inspiration, usually as a result of an individual visualising how much better his/her working life and his/her company could be and should be. Initial progress along the road to TPM is slow and difficult, but through hard work the momentum of change gradually increases until it is unstoppable. The road never ends, because TPM is an ongoing way of life, but it does become less steep as the true philosophy of TPM takes hold and changes the culture of the whole company.

Many companies have just started along the road to TPM and are finding that the gradient is very difficult to overcome, and many more are wondering just how to start the journey. I spend a great deal of my time helping operating companies to start and supporting them along the road to TPM, and everyone finds it not only difficult at times but also very rewarding. The key is to keep going and be determined to make it happen, and eventually the gradient will reduce.

TPM is applied common sense, in fact, to run an operating company without the TPM philosophy and practices in place is so wasteful and plainly wrong that it can be difficult to believe that so many companies still work that way. The benefits of TPM are there for everyone in the company and its implementation is one of the rare 'win-win' scenarios that I have encountered in industry. TPM is a process of change – change in the values of a company, change in the relationships between people, in the way in which a company operates and in the very culture of a company. To implement this degree of change is difficult and the timescale is measured in years, and during TPM implementation there will be occasions when the pace of change slows down or even seems to stop. At times like these, it is easy to get downhearted and ask 'why are we doing this?', but take comfort from the fact that anyone involved with change will tell you that the road to change is littered with obstacles and that every company undergoing change will, at some point feel that it is failing. Companies that then abandon the programme have failed, but those that renew their efforts and persevere with what they know is right, inevitably succeed. It is important to be tenacious and to hold on to the vision of what the company can and will be like.

I believe that the cause of TPM is both noble and righteous. It is about respect for each other, our company and ourselves; it is about helping each other and working together as a team; it is about self-discipline; it is about transforming the way in which a company operates; it is about the enrichment of people's lives. I sincerely hope that this book will encourage the reader to promote the cause of TPM in whatever organisation he/she is employed and I hope that my enthusiasm and belief will inspire you to become one of the many TPM champions who are needed to transform our operating companies.

TPM in action!

Appendix 1

Company Checklist

■ Audit of machine condition and attitudes within the area

Company...

Department... Date ....

Tick the box which most closely describes the conditions, situation and attitudes prevailing.

Machinery condition

The machinery and equipment is generally dirty.	Yes ☐	No ☐
Swarf is scattered on and around machinery.	Yes ☐	No ☐
Cutting oil is splashed on and around machinery.	Yes ☐	No ☐
The machinery leaks hydraulic fluid.	Yes ☐	No ☐
The machinery leaks lubricant.	Yes ☐	No ☐
Oil pans are full, often to overflowing.	Yes ☐	No ☐
Motors are coated with a layer of oil/grime.	Yes ☐	No ☐
Grime from cutters and grinders is caked on to the machinery.	Yes ☐	No ☐
Limit switches on machinery are covered with oil/grime.	Yes ☐	No ☐
Covers are used to protect certain machinery or machine areas but their internal parts are not cleaned or inspected.	Yes ☐	No ☐
Some machinery parts rattle and vibrate.	Yes ☐	No ☐
Machinery is positioned so that access for routine maintenance is difficult.	Yes ☐	No ☐
Oil cans are left around, often empty and dirty.	Yes ☐	No ☐
Drains and filters are clogged.	Yes ☐	No ☐

Wires and pipes are left in an untidy mess, making it hard to tell which goes where. Yes ☐ No ☐
People do not mind seeing dirt, process waste, swarf and oil pile up on machinery, as they think it is normal. Yes ☐ No ☐
Motors are allowed to get hot or make strange noises without it being detected. Yes ☐ No ☐
Quick fixes are often put in place on machinery such as parts being wired up or the correct number of vee belts not being replaced. Yes ☐ No ☐

Number of Yes ☐ Number of No ☐

Comments by auditor:
....
....

Condition of area around machinery

Swarf is scattered around and has to be swept up. Yes ☐ No ☐
The floor is dirty and, in places, slippery with oil. Yes ☐ No ☐
Jigs and tools are untidy and left lying around. Yes ☐ No ☐
There are a lot of useless items lying around. Yes ☐ No ☐
Tools, materials, etc. are not kept in specified places. Yes ☐ No ☐
There is no specified place which is clearly marked for quality inspection equipment to be kept. Yes ☐ No ☐
Wipers, paper cups and cigarette packets are lying around. Yes ☐ No ☐
There are no stands/racks for oil cans and equipment. Yes ☐ No ☐

Number of Yes ☐ Number of No ☐

Comments by auditor:
....
....

Machinery operators

Operators do not carry out regular machinery inspections, as they probably do not know how to. Yes ☐ No ☐
Only some operators know where, when and how to oil their machinery and they do not always carry it out correctly. Yes ☐ No ☐
When operators find a machinery fault they call for maintenance support without trying to fix it themselves. Yes ☐ No ☐
Operators do not regard breakdowns and product defects as their problem. Yes ☐ No ☐

Operators do not know how to carry out simple machinery repairs. Yes ☐ No ☐
Operators do not carry out quality checks. Yes ☐ No ☐
Operators sometimes use equipment and instruments that are inaccurate or defective. Yes ☐ No ☐

Number of Yes ☐ Number of No ☐

Comments by auditor:

....

....

General conditions and performance

Machinery breakdowns occur quite frequently, at a rate of 3 per cent of operating time or higher. Yes ☐ No ☐
Repairs generally take a long time to complete. Yes ☐ No ☐
Minor problems occur quite regularly and often the repair is only temporary. Yes ☐ No ☐
Breakdowns occur for the same reasons, time and time again. Yes ☐ No ☐
Changeover and set-up adjustments take a lot of time. Yes ☐ No ☐
People accept ongoing adjustments as normal. Yes ☐ No ☐
Problems following changeover occur more (or less) often, depending upon who does it. Yes ☐ No ☐
Minor stoppages happen very often. Automated machinery often needs a minder. Yes ☐ No ☐
Re-working occurs at a rate of 3 per cent or more. Yes ☐ No ☐
Scrap from machinery is running at 2 per cent or more. Yes ☐ No ☐
Machinery speeds have been decreased to reduce scrap and/or wear. Yes ☐ No ☐
Product specific standard cycle times have not been established. Yes ☐ No ☐
Operators know the standard times but do not keep to them. Yes ☐ No ☐
No one has analysed speed losses in the machinery. Yes ☐ No ☐
There are no charts showing how effective the machinery is. Yes ☐ No ☐
There are no procedures for cleaning, lubricating, etc. Yes ☐ No ☐

Number of Yes ☐ Number of No ☐

Comments by auditor:

....

....

Total number of Yes ☐ Total number of No ☐

Interpretation of the results

Total Number of Yes	Comments
40+	Machinery condition and attitudes are very poor. Operating performance is well below par and morale is suffering. Immediate action is needed.
30–40	Poor machinery condition and bad attitudes are seriously affecting your operating performance. You need to address this urgently.
20–30	You have obviously understood the importance of machinery condition and taken some action. More progress can be made to significantly improve your operating performance.
10–20	Significant progress has been made or perhaps the emphasis has always been on keeping machinery in good condition. Keep up the good work and address the 'Yes' problems.
< 10	Well done! You are well on the way to top-class performance. Work towards zero.

Comments by auditor:

....

....

....

....

....

Appendix 2
Machinery Design Checklist

■ Audit of machinery design and manufacture for TPM

Company...

Department.. Date ....

Tick the box which most closely describes the general features of the machinery.

Machine structure

Process waste is restricted to the processing area.	Yes ☐	No ☐
Extraction/filtration provided for mist or fumes.	Yes ☐	No ☐
Adequate inspection windows to view the process.	Yes ☐	No ☐
Means of cleaning the process area provided.	Yes ☐	No ☐
Sharp corners and pockets either eliminated, filled in or covered.	Yes ☐	No ☐
Covers easily removed (handles, quick release fasteners, etc.).	Yes ☐	No ☐
Easy access to main machine elements is provided.	Yes ☐	No ☐
Moving parts are guarded and interlocked (where required).	Yes ☐	No ☐
Covers are provided around components that may get contaminated.	Yes ☐	No ☐
Covers/catchers are provided around components that may deposit oil or grease.	Yes ☐	No ☐

Mechanical components

Moving parts are clearly identified. Yes ☐ No ☐
Gearbox oil site levels are easily viewed and marked with red and green paint. Yes ☐ No ☐
Gearbox oil filler caps are easily accessible, clearly identified and the type of lubricant is indicated. Yes ☐ No ☐
Return/collection systems are provided for lubricants. Yes ☐ No ☐
Centralised greasing systems are provided. Yes ☐ No ☐
Means of gearbox removal/replacement are provided, i.e. access and handling are not difficult. Yes ☐ No ☐
Access for main bearing removal/replacement is not difficult. Yes ☐ No ☐
Condition monitoring is provided for main drives, i.e. thermal tape, vibration nodes, etc. Yes ☐ No ☐
Vibration resistant fasteners and/or matchmarks are used. Yes ☐ No ☐
Setting marks/instructions are provided on adjustment components. Yes ☐ No ☐

Waste removal

Process waste is directed away from working parts. Yes ☐ No ☐
There are no sharp corners, pockets or trapping points in the processing or waste removal area. Yes ☐ No ☐
The waste disposal path is leakproof. Yes ☐ No ☐
The waste removal rate and container size are adequate. Yes ☐ No ☐
Waste bin removal and transportation are easy. Yes ☐ No ☐
Means of catching any waste material at product output is provided. Yes ☐ No ☐
Means of catching any waste material during changeover is provided. Yes ☐ No ☐

Hydraulics/pneumatics and other fluid systems

Pressure gauges are easily viewed. Yes ☐ No ☐
Pressure gauges are marked with red and green. Yes ☐ No ☐
Manual valves are colour-coded to indicate their normal position. Yes ☐ No ☐
Direction of and type of fluid flow are indicated on pipes. Yes ☐ No ☐
Fluid site levels are easily viewed and marked with red and green paint. Yes ☐ No ☐
Filler caps are easily accessible, clearly identified and the type of fluid is indicated. Yes ☐ No ☐

Filters are easily accessible and clearly marked.	Yes ☐	No ☐
Differential pressure gauges are used where possible to indicate filter condition.	Yes ☐	No ☐
Drain plugs are easily accessible and clearly marked.	Yes ☐	No ☐
Space is provided for a receptacle or pipe to drain the fluid.	Yes ☐	No ☐
Fluid sprays are adjustable and fitted with anti-surge valves.	Yes ☐	No ☐
Pipe runs are neat and tidy.	Yes ☐	No ☐
Pipe runs do not form dirt traps and pockets.	Yes ☐	No ☐
Access is provided for cylinder removal and replacement.	Yes ☐	No ☐

Electrical controls

Cable runs are neat and tidy.	Yes ☐	No ☐
Cable runs do not cause dirt traps and pockets.	Yes ☐	No ☐
Cable runs are not over or in front of covers.	Yes ☐	No ☐
Wires are identified – numbered or colour-coded.	Yes ☐	No ☐
There is easy access for wire removal and replacement, with probes and tightening screws at terminal fields.	Yes ☐	No ☐
All control cabinets are fully sealed.	Yes ☐	No ☐
All control cabinets are secure.	Yes ☐	No ☐
Control cabinets are clearly identified.	Yes ☐	No ☐
Meters are easily viewed and marked with red and green.	Yes ☐	No ☐
Operator controls are clearly marked.	Yes ☐	No ☐
Operator controls are easily accessible during operation.	Yes ☐	No ☐
Switches and sensors are clearly marked.	Yes ☐	No ☐
Switches and sensors are solidly mounted.	Yes ☐	No ☐
Switches and sensors are guarded where necessary.	Yes ☐	No ☐
Setting marks and/or instructions are provided for adjustable switches/sensors.	Yes ☐	No ☐
Control software contains 'safe home' re-set routines.	Yes ☐	No ☐
Control software provides simple diagnostics.	Yes ☐	No ☐

Total number of Yes ☐ Total number of No ☐

Comments by auditor and any recommendations:

....

....

....

....

....

....

....

Failure Modes and Effects Analysis (FMEA)											
System			Assembly				Component				
No.	Part/ Function	Failure Mode	Poss. Causes	Fail Detection	Available Counters	Fail Effects	Occurrence	Severity	Detection	RPN	Failure Assessment Actions

Appendix 3

Overall Effectiveness Examples and Exercises

Overall Effectiveness Examples and Exercises

Example 1

planned time to run = 8 hours (480 minutes)

breaks and scheduled maintenance = 20 minutes

loading time = 480 − 20 = 460 minutes

down-time encountered = 20 minutes breakdowns and 40 minutes changeover and adjustment

$$\% \text{ availability} = \frac{\text{loading time} - \text{down-time}}{\text{loading time}} = \frac{460 - 20 - 40}{460} \times 100 = 87\%$$

output = 400 components

theoretical cycle time = 0.5 minutes/component

$$\% \text{ performance} = \frac{\text{actual output}}{\text{potential output}} = \frac{400}{(400/0.5)} \times 100 = 50\%$$

number of reject components = 8

$$\% \text{ quality} = \frac{\text{produced units} - \text{defects}}{\text{produced units}} = \frac{400 - 8}{400} \times 100 = 98\%$$

overall effectiveness = 0.87 × 0.5 × 0.98 × 100 = 42.6%

Example 2

planned time to run = 120 hours/week (7200 minutes)

breaks and scheduled maintenance = 0 (continuous production) loading time = 7200 − 0 = 7200 minutes

down-time encountered = 120 minutes breakdowns and 460 minutes changeover and adjustment

$$\% \text{ availability} = \frac{\text{loading time} - \text{down-time}}{\text{loading time}} = \frac{7200 - 120 - 460}{7200} \times 100$$

$$= 92\%$$

output = 220 000 units

theoretical output = 2250 components/hour

$$\% \text{ performance} = \frac{\text{actual output}}{\text{potential output}} = \frac{220\,000}{(6620/60 \times 2250)} \times 100 = 88\%$$

number of reject units = 2050

$$\% \text{ quality} = \frac{\text{produced units} - \text{defects}}{\text{produced units}} = \frac{220\,000 - 2050}{220\,000} = 99\%$$

overall effectiveness = 0.92 × 0.88 × 0.99 × 100 = 80%

Exercise 1

planned time to run = 80 hours (4800 minutes)

breaks and scheduled maintenance = 210 minutes

loading time = 4800 − 210 = 4590 minutes

down-time encountered = 400 minutes breakdowns and 960 minutes changeover and adjustment

% availability = ________________________ × 100 = %

output = 75 300

theoretical output = 1500 units/hour

% performance = ________________________ × 100 = %

number of reject units = 1100

% quality = ________________________ × 100 = %

overall effectiveness = × × × 100 = %

Exercise 2

planned time to run = 144 hours/week (8640 minutes)

breaks and scheduled maintenance = 0 (continuous production)

loading time = 8640 − 0 = 8640 minutes

down-time encountered = 700 minutes breakdowns and 480 minutes changeover and adjustment

% availability = ________________________ × 100 = %

output = 57 tonnes

theoretical output = 0.6 tonnes/hour

% performance = ________________________ × 100 = %

amount of waste product = 0.8 tonnes

% quality = ________________________ × 100 = %

overall effectiveness = × × × 100 = %

Exercise 1 – Solution

planned time to run = 80 hours (4800 minutes)

breaks and scheduled maintenance = 210 minutes

loading time = 4800 − 210 = 4590 minutes

down-time encountered = 400 minutes breakdowns and 960 minutes changeover and adjustment

$$\% \text{ availability} = \frac{4590 - 400 - 960}{4590} \times 100 = \frac{3230}{4590} \times 100 = 70\%$$

output = 75 300 units

theoretical output = 1500 units/hour

$$\% \text{ performance} = \frac{75\,300}{(3230/60) \times 1500} \times 100 = \frac{75\,300}{80\,750} \times 100 = 93.2\%$$

number of reject units = 1100

$$\% \text{ quality} = \frac{75\,300 - 1100}{57\,300} \times 100 = \frac{74\,200}{75\,300} \times 100 = 98.5\%$$

overall effectiveness = 0.70 × 0.932 × 0.985 × 100 = 64.3%

Exercise 2 – Solution

planned time to run = 144 hours/week (8640 minutes)

breaks and scheduled maintenance = 0 (continuous production)

loading time = 8640 − 0 = 8640 minutes

down-time encountered = 700 minutes breakdowns and 480 minutes changeover and adjustment

$$\% \text{ availability} = \frac{8640 - 700 - 480}{8640} \times 100 = \frac{7460}{8640} = 86\%$$

output = 57 tonnes

theoretical output = 0.6 tonnes/hour

$$\% \text{ performance} = \frac{57}{(7460/60) \times 0.6} \times 100 = \frac{57}{74.6} = 76\%$$

amount of waste product = 0.8 tonnes

$$\% \text{ quality} = \frac{57 - 0.8}{57} \times 100 = \frac{56.2}{57} = 98.6\%$$

overall effectiveness = $0.86 \times 0.76 \times 0.986 \times 100 = 64\%$

The cost of the six big losses

Example 1

theoretical output = 2000 units/hour

normal running hours = 120 hours/week, 46 weeks/year

value added by the machinery = £0.02/unit

OE of machinery = 63% average

therefore:

theoretical earning capacity of the machinery = £0.02 × 2000 × 120 × 46
= £220 800/year

actual earning capacity = £220 800 × 0.63
= £119 104/year

the six big losses cost: £220 800 − £119 104 = £81 696/year

$$\text{increased earning capacity per 1\% improvement in OE} = \frac{£81\,696}{37}$$

$$= £2208/1\%$$

If this was an average figure and the company had 50 machinery items, and if each of them could be improved by 10 per cent (which is quickly achieved through TPM), then the increase in earning capacity for the company would be:

£2208 × 10 × 50 = £1.1 million

for no additional manufacturing costs.

Example 2

theoretical output = 70 units/hour

normal running hours = 75 hours/week, 46 weeks/year

value added by the machinery = £1.50/unit

OE of machinery = 55% average

therefore:

theoretical earning capacity of the machinery = £1.50 × 70 × 75 × 46
= £362 250/year

actual earning capacity = £362 250 × 0.55
= £119 238/year

the six big losses cost: £362 250 − £199 238 = £163 012/year

increased earning capacity per 1% improvement in OE $= \dfrac{£163\,012}{45}$
= £3622/1%

If this was an average figure and the company had 50 key machinery items, and if each of them could be improved by 10 per cent (which is quickly achieved through TPM), then the increase in earning capacity for the company would be:

£3622 × 10 × 50 = £1.8 million

for no additional manufacturing costs.

Appendix 4

Six Big Losses

Data Collection Form

The Six Big Losses

Does the machinery ever break down?	Yes ☐ No ☐	
If so, how often and why?		Average breakdown time per shift/day
Usually for how long?		
Does the machinery have to be changed over and/or set up?	Yes ☐ No ☐	
If so, how often?		Average c/over time per shift/day
How long does it normally take?		
Does the operator have to stop due to: • minor breakdowns which are fixed quickly? • things that stick? • the machinery having to be shutdown or re-set?	Yes ☐ No ☐	
If so, how often?		Average stoppage time per shift/day
How long does each stoppage last, on average?		
Is the machinery/process working more slowly than it should?	Yes ☐ No ☐	
How often does this happen?		Average lost prodn. per shift/day
How does this affect the production rate?		
Does the machinery/process ever produce scrap or product that has to be re-processed?	Yes ☐ No ☐	
If so, how often?		Average loss per shift/day
How much product or time is lost on average?		

Total Productive Maintenance

The TPM Centre

Appendix 5

Overall Effectiveness

Total Productive Maintenance – Machine Data Collection Sheet

Machine .... Cell/Area..............................

Shift Date & Time	Job	Loading Time			Availability			Performance		Quality
	Name or No.	Start Prod	Stop Prod	Total Time	C/Over & Set Up	B/Down No.	B/Down Time	Stoppages Total Time	Slowdown Total Time	Scrap or Re-work Time Lost
Totals										

Total Productive Maintenance – Machine Data Collection Sheet

Machine .... Cell/Area..............................

Shift Date & Time	Job	Loading Time			Availability			Performance			Quality	
	Name or No.	Start prod	Stop prod	Total Time	C/Over & Set Up	B/Down		No. Made	Std Time per Part	Stop Time	No. Scrap	Re-work No./ Time
						No.	Time					
Totals												
Totals												

Total Productive Maintenance – Machine Data Collection Sheet

Machine ......H&K1...... Cell/Area...F.C. UNITS...

Shift Date & Time	Job	Loading Time			Availability			Performance			Quality	
	Name or No.	Start prod	Stop prod	Total Time	C/Over & Set Up	B/Down No.	B/Down Time	No. Made	Std Time per Part	Stop Time	No. Scrap	Re-work No./ Time
11/4/94	DL7	7·30	4·00	8	1½	1	½	310	1 MIN.	–	10	2
12/4/94	DL7	7·30	12·00	4½	–	1	1	200	1 MIN	–	5	–
	GD12	1·00	4·00	3	2	–	–	110	½ MIN	–	8	2
13/4/94	GD12	8·00	2·00	5½	–	2	½	540	½ MIN	–	24	–
	LD4	2·00	4·00	2	1	–	–	50	1 MIN	–	–	–
14/4/94	LD4	7·30	4·00	8.	–	2	2	290	1 MIN	–	12	6.
15/4/94	LD4	8·00	12·00	4	–	1	½	200	1 MIN		4	–
Totals				35	4½	7	4½	1700		–	63	10
Totals												

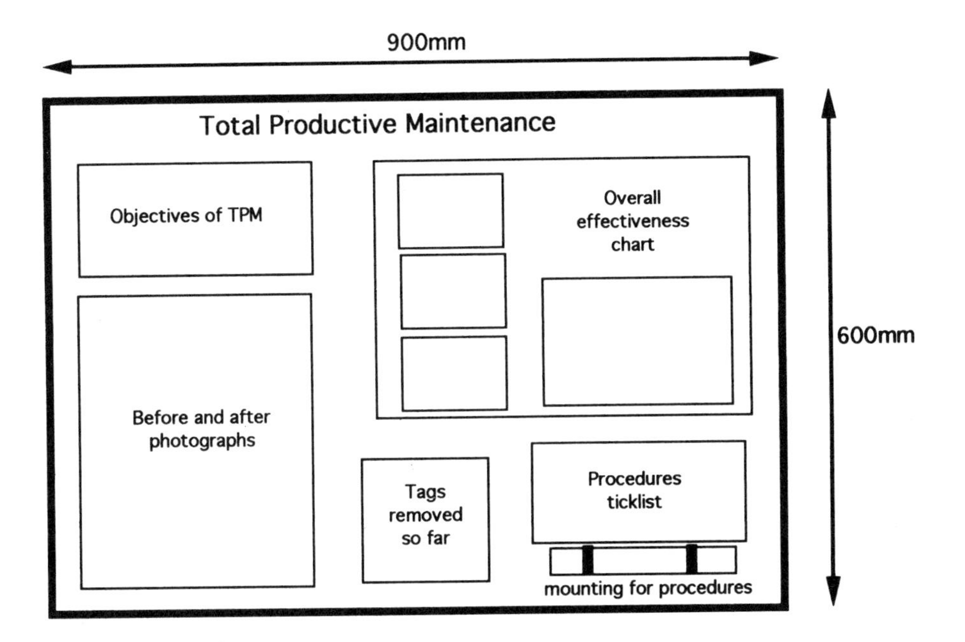

TPM station for mounting an overall effectiveness chart

Appendix 6

Maintenance Procedure

Autonomous Maintenance Procedures

Area;	Equipment;	Sheet of	Issue Date;

No.	Maintenance Activity	What to Use/Notes	Interval		By
			Week	Month	

Safety Note

Total Productive Maintenance

The TPM Centre

Cleaning Procedures

Area; xxxxxxxxx	Machine; xxxxxxxxxxxxxxxx	Sheet 1 of 1	Issue Date; 4/10/93

No.	Where and What to Clean	What to Use	Interval			By
			Day	Week	Month	
1	clean magnetic table surface	coolant hose	●			Op
2	clean grinding dust from table	coolant hose		●		Op
3	wipe over front control panel	wiper + detergent if necessary		●		Op
4	wipe oil drips from head traverse slideways	wiper + detergent if necessary		●		Op
5	clean around machine surfaces and floor	wiper + detergent if necessary + broom/vacuum			●	Op/ Con.
6	clean under magnet and re-grind if necessary	brush/vacuum (remove magnet)			● 12	Op

Safety Note

Make sure that the machine is turned off and the power isolated when cleaning the grinding table and head .

Total Productive Maintenance

The TPM Centre

Bolt Tightening and Inspection Procedures

Area; xxxxxxxx	Machine; xxxxxxxxxxxxxxx	Sheet 1 of 1	Issue Date; 4/10/93

No.	Location	Operation/Inspection	Tool to Use	Interval			By
				Day	Week	Month	
1	coolant tank	check coolant level and top up		●			Op/lab
2	coolant tank	drain, clean and re-fill coolant tank	vacuum			●	Maint
3	coolant tank	check filter paper - renew roll as reqd.		●			Op.
4	table drive	check drive belts and pulleys	spanners to remove cover		●		Op.
5	rear of m/c	check/replace air filter on main column				●	Op
6	coolant system	run coolant system to stop stagnation	*when m/c is not in use*	●			Op

Safety Note Make sure that the machine is turned off and power isolated when checking the table drive and changing coolant.

Total Productive Maintenance

The TPM Centre

Lubrication Procedures

Area; xxxxxxxxx	Machine; xxxxxxxxxxxxxxxxx	Sheet 1 of 1	Issue Date; 4/10/93

No.	What to Lubricate	Lubricant to Use	Any Special Tool	Interval			By
				Day	Week	Month	
1	check main spindle drive fluid level	kerosene			●		Op
2	check hydraulic oil level	hydraulic oil				●	Op
3	check oil level in coolant system gearbox	SAE 90				●	Maint.
4	check and grease bearings, chains and sprockets on coolant system drive	grease	grease gun			● 3	Maint.
5	grease main table drive	grease	grease gun		●		Maint.

Safety Note

Make sure that the machine is turned off and power isolated when checking or greasing the coolant system drive.

Total Productive Maintenance

The TPM Centre

Total Productive Maintenance

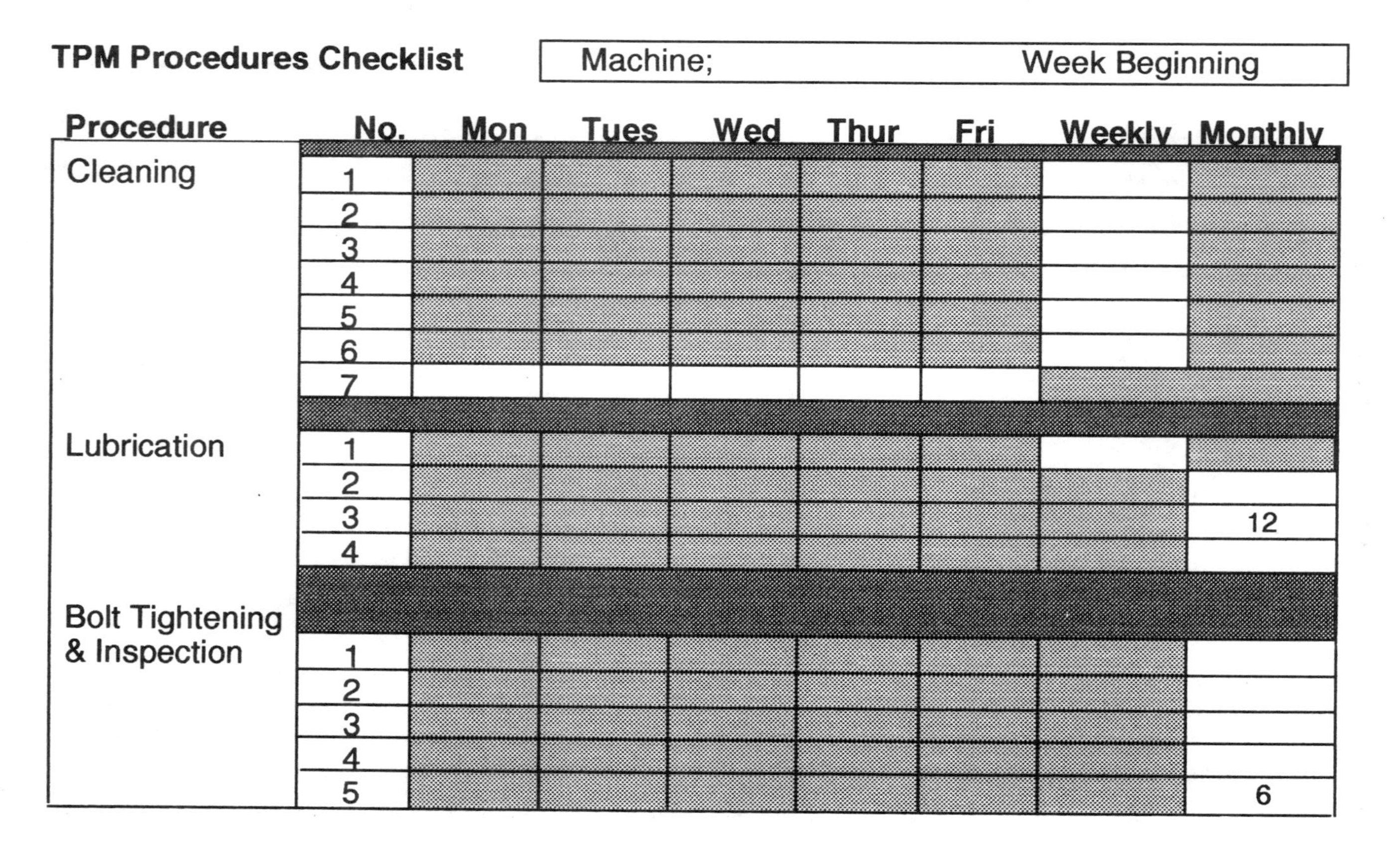

TPM Procedures Checklist

Machine;	Week Beginning

Procedure	No.	Mon	Tues	Wed	Thur	Fri	Weekly	Monthly
Cleaning	1							
	2							
	3							
	4							
	5							
	6							
	7							
Lubrication	1							
	2							
	3							12
	4							
Bolt Tightening & Inspection	1							
	2							
	3							
	4							
	5							6

Index